THE CURE FOR EVERYTHING!

THE CURE FOR EVERYTHING!

Untangling the Twisted Messages About Health, Fitness, and Happiness

TIMOTHY CAULFIELD

VIKING

VIKING
an imprint of Penguin Canada

Published by the Penguin Group
Penguin Group (Canada), 90 Eglinton Avenue East, Suite 700, Toronto, Ontario, Canada M4P 2Y3
(a division of Pearson Canada Inc.)

Penguin Group (USA) Inc., 375 Hudson Street, New York, New York 10014, U.S.A.
Penguin Books Ltd, 80 Strand, London WC2R 0RL, England
Penguin Ireland, 25 St Stephen's Green, Dublin 2, Ireland (a division of Penguin Books Ltd)
Penguin Group (Australia), 250 Camberwell Road, Camberwell, Victoria 3124, Australia
(a division of Pearson Australia Group Pty Ltd)
Penguin Books India Pvt Ltd, 11 Community Centre, Panchsheel Park, New Delhi – 110 017, India
Penguin Group (NZ), 67 Apollo Drive, Rosedale, Auckland 0632, New Zealand
(a division of Pearson New Zealand Ltd)
Penguin Books (South Africa) (Pty) Ltd, 24 Sturdee Avenue, Rosebank,
Johannesburg 2196, South Africa

Penguin Books Ltd, Registered Offices: 80 Strand, London WC2R 0RL, England

First published 2012

1 2 3 4 5 6 7 8 9 10 (RRD)

Manufactured in the U.S.A.

LIBRARY AND ARCHIVES CANADA CATALOGUING IN PUBLICATION

Caulfield, Timothy A., 1963-
The cure for everything! : untangling the twisted messages about health, fitness, and happiness /
Timothy Caulfield.

Includes index.
ISBN 978-0-670-06523-3

1. Health—Popular works. 2. Exercise—Popular works.
3. Happiness. 4. Nutrition—Popular works. 5. Medicine, Popular.
6. Alternative medicine—Popular works. I. Title.

RA776.C38 2012 613 C2011-906594-0

Visit the Penguin Canada website at **www.penguin.ca**

Special and corporate bulk purchase rates available; please see
www.penguin.ca/corporatesales or call 1-800-810-3104, ext. 2477.

FOR MY WORKOUT PARTNER, JOANNE

CONTENTS

INTRODUCTION
THE COOKIE CONUNDRUM

"Ninety-five percent of all disease is caused by a build-up of acid in our bodies," said the instructor emphatically. She had an unusual accent. A mix of Bavarian barmaid and Southern preacher. She was large and sturdy, and seemed to believe everything she was saying. Every seat in the room was taken by her audience of middle-aged couples, so I sat on some exercise equipment at the back. I was on summer holiday and my wife, Joanne, had convinced me to attend the lecture. "I bet it will be fun," she said. "Who knows? You might learn something." Her smile told me she knew I wasn't likely to agree with her optimistic knowledge-enhancement prediction.

"We need to cleanse our bodies," the instructor said. "Just like a car, we need to clean our bodies to make sure they work properly so we don't get diseases like cancer and diabetes. By cleansing our bodies we will become more fit and our metabolism will increase and we will lose weight, especially in all those tough spots: the arms, the belly, and the bum." She pointed to the relevant region of the body as she said *arms, belly*, and *bum*. The latter got a laugh from the crowd. They were captivated. Some were writing notes as she explained the details of detoxification.

I knew that absolutely every statement she made was incorrect or misleading. Complete crap. We don't need to detoxify and cleanse our bodies. Detoxification will not result in weight loss. You cannot cause a particular part of your body to lose weight by consuming algae, which was her recommended remedy for the elimination of flab.

It was, however, an amazing performance, a mix of pseudo-scientific jargon, faith-healing proselytization, and over-the-top fear mongering. Some of her claims were so absurd that I had to suppress a laugh. For example, she informed the crowd that if you eat meat and don't "cleanse" on a regular basis (i.e., irrigate your bowel) the lower colon gets clogged with an immovable, thick, mucousy sludge. This sludge, apparently, causes a host of ailments, and some of us are carrying around as much as ten pounds of this disgusting substance. Some of her other claims infuriated me because of their simplistic inaccuracies, especially those that implicated a serious disease. She told the audience, her grave tone reinforced by scientific-looking diagrams and terminology, that the toxins in deodorants cause breast cancer. Nonsense.

At the end of the talk came the sales pitch. "Do you want to be healthy? You can't blame aging," the instructor informed us all. "You can only blame yourself. You must take control and detoxify your body. An acidity and toxicity assessment costs $40. Isn't your health worth this much?"

Coincidentally, a host of detoxifying products and processes would be available for sale after the assessment. As soon as the talk ended, it appeared that everyone in the room lined up to get detoxified.

And who can blame them? Despite my deep skepticism, I understood the urge to jump to the head of the line. Who wants sludge in their colon? Detoxifying sounds like a sensible idea. Don't toxins cause disease? And while the fat-melting algae potion was a bit far-fetched, much of her presentation seemed grounded in some kind of science. She was offering a path to better health.

And I want to be healthy.

This book is about health. More particularly, it is about the science associated with health. As the detoxifying dominatrix demonstrated, there is a considerable amount of weird information out there. Every day we are showered with advice about our health. We are told we are fat. We are told what to eat. We are told what not to eat. We are told to cleanse. We are told to take supplements. We are told we need to exercise. We are told to stretch. We are told to take pharmaceuticals. We are told to avoid pharmaceuticals. We are told to get our meridians centred.

We are told to get healthy, damn it!

Of course, it's true that there has never been more evidence—of vastly differing degrees of quality—to support this advice. We live in the health-science era. It's everywhere. Pick up a newspaper: there is a good chance you will find several stories on biomedical research. Often the stories are frivolous reports describing a new diet or exercise routine. But some others may be about the latest big science discoveries—a scientific breakthrough, for instance, promising life-enhancing remedies in just a few years. There are also stories that come from the scientific fringe. My local newspaper seems to have at least one article a week on the benefits of some form of alternative remedy, practices deriving from the realms of homeopathy, chiropractic, and naturopathy. Regardless of the source, all of these stories have a common theme: our health.

Newspapers, of course, are hardly our only source of health and science stories. An ever-rising tide of information can be found on television, on radio, in government reports, on blogs, on iPhone apps, in books and magazines, and, of course, in advertising about pharmaceuticals, diets, and fitness programs. We live in a sea of (purportedly) science-based health information.

What are we to do, really, with all this information? Can we actually use it to live a healthier life? What information can we trust? Are emerging areas like genetics the answer to our health problems? Do any diets work? Can we ever believe pharmaceutical companies? Is the simple answer really that we all just need to clean our colons? Is there a cure for everything?

In this book I will seek to answer these and other related questions. What will we find? While science is everywhere, the scientific information that passes through our field of view is often wrong, hyped, or twisted by an ideological or commercial agenda. It may also be influenced by our own beliefs, evolutionarily determined predilections, or market-leveraged desires.

We need to recognize the deep irony of it all. At a time when scientific knowledge has never been more important, it is being subjected to an unprecedented number of perverting influences. Not that this should come as a surprise. As science becomes more central to our lives, the stakes grow higher and the incentives to twist the scientific message multiply. In some countries, health care consumes from one-third to one-half of government spending. Both the pharmaceutical and alternative-medicine industries generate hundreds of billions of dollars in annual sales. And many billions of dollars have been invested in research initiatives such as the Human Genome Project.

The result is clear: the health sciences are more essential to our world than ever before. And their impact—on the economy, on human health, on education, and on the broader culture—is likely to increase. Science informs the decisions we make about the drugs we take, the food we eat, and the forms of exercise we take up. Scientific data should help us make the best decisions we can, but is this what actually happens? While never promising the Truth, science is meant to nudge us closer to an objective picture of our world. It should give us some idea of what is likely to work. But despite the power of science to provide us with practical information, most of us have only a vague notion of what actually makes us healthy. Our vision has been obscured by personal and institutional agendas, commercial interests, and pop-culture spin. And, to make matters worse, the institutions that should guard us against these perverting influences have mostly failed at the task. In fact, because governments view scientific innovation as the engine of economic growth, they have become part of the problem. They, too, have become twisted.

My aim is to clarify the picture and explain the forces that have made it obscure. I will try to strip away the hype and the agendas to reveal the sound evidence where it exists, and its absence when the research hasn't been done. I will also provide practical advice (and warnings) for those wishing to navigate the churning sea of facts, findings, and fears associated with emerging health technologies and health-promoting strategies.

The road to good health is simpler than we are often led to believe. In some ways, this is liberating. Ninety percent of a healthy lifestyle is associated with a few simple truths. It is not necessarily an easy path to follow, but if you can parse the twisted messages that bombard us daily, you'll find that the way is surprisingly direct.

Many scientists believe that if people just knew more about science or understood the facts, they would be more rational about their health decisions. This view, which has been called the "deficit model," is faulty. Research has shown that learning about science *can* have a dramatic impact on an individual's view about a health issue—an impact that I optimistically and perhaps naively hope this book will have—but that this is not the norm. In fact, supplying individuals with scientific facts rarely alters beliefs. People see, select, and interpret information about health (and many other topics) through an individual and largely self-constructed lens of preconceived beliefs, values, and fears. We all, including many researchers, typically behave like cognitive contortionists in order to find ways to keep believing what we're predisposed to believe. And, of course, the facts themselves are often either disputed or twisted beyond recognition. As I will show, the scientific community producing the facts is often responsible for a significant part of the distortion.

The facts are all too frequently twisted the moment they leave the lab.

Given these realities, I felt it was essential to do two things in this book. First, I wanted to *experience* this journey through the world of health science. I wanted to do more than merely speak to experts and read the relevant research (though this kind of traditional scholarship remains at the heart of my exploration). I felt it important to get some real context. I needed to actually go to the folks who are providing health-care products and advice—researchers, MDs, alternative-medicine practitioners—to appreciate in person what people are hearing, seeing, and feeling. And so, for each health-science area I investigated, I did my level best to get involved. For the fitness chapter, I went to see a renowned personal trainer. For the diet chapter, I went on a diet. For the genetics chapter, I got my genes tested. And for the remedy chapter, I tried a host of potions and procedures.

It would be wrong for me to assert that these limited experiences gave me a rich appreciation of other people's world views. But without exception my experiences provided me with a new and often enlightening perspective on each topic. This frankly came as a surprise. As a professor of health law and policy, I have spent the last 18 years of my career studying health-policy issues. One of the perks of my job (well, it's a perk if you love looking at data) is the opportunity this has given me to spend much of my day reviewing the relevant research. I have also conducted research of my own on the topics covered in this book. I have, for example, been a principal investigator on several large projects exploring the health and policy issues associated with genetics. I have also investigated obesity policy and the legal issues associated with the provision of alternative medicine. In addition, I have worked with many of the world's leading scientists and sat on innumerable national and international health- and science-policy committees. But despite a life immersed in the issues, the personal experience I acquired in the course of researching this book provided a new and invaluable dimension to my analysis. Also, there was a bit of pride involved. If I was going to talk the talk, I thought I had better be prepared to walk the walk, eat the broccoli, and take the medicine.

The second goal I set for myself was to cast my evidence net very broadly. One of the academic experts I interviewed for the fitness chapter noted that we often get things wrong because we look at health issues through the lens of a single discipline. But, she told me, "the factors affecting health are so complex and diverse that you need to look at it from the perspective of a wide range of scientific disciplines. If you don't do this, you'll see everything only in terms of your discipline." In other words, a geneticist sees things in terms of genetics. A nutritionist sees things in terms of diet. If the only tool you've got is a hammer, goes the old saying, then pretty soon everything starts to look like a nail. I have tried to use more than one tool.

I would like to make several disclaimers before we start our journey.

The first is that I recognize that not everyone places being healthy at the centre of their universe. Not everyone is obsessed with the maximization of health. I was able to reflect upon this reality while writing this introduction. I was working (you'll have to take my word for this) while camped out at a table in a popular pub in downtown Toronto. Just before I closed my laptop to call it a night, I struck up a conversation with the two twenty-something guys sitting next to me. All night they had been laughing, gesticulating, and drinking pint after pint. They seemed like good-natured fellows, fast friends, and they were having a great time. I asked them how they met.

"We met at a gym. I was working on this," said one, as he grabbed a handful of belly. "We were working out and we both went for a smoke break at the same time. We've been buddies ever since."

"There are gyms in this city that let you smoke?"

"No," one of them replied. "We went outside."

I pictured these two not very sleek individuals standing on a sidewalk in their gym clothes, happily smoking away.

"Fucking nanny state!" opined the other. Fist bump.

Indeed. What a grave injustice it is that you can't smoke while jogging on a treadmill at your local gym! These fellows struck me as not being particularly concerned with their health, at least not at that moment in their lives. They might want to slim down a bit—if for no other reason than to increase their chances of catching the attention of the finer sex—but given their impressive beer consumption, I suspect their determination to lose weight waned from time to time.

But the truth is that we are all going to die. Yep. Given this inevitability, the constant battle to stay healthy and extend our lives is a tad irrational. Why not simply embrace life? Have fun. Smoke a cigarette on the Stairmaster. Drink beer with your pals. The glass, as the Buddhist saying goes, is already broken.

I don't dismiss this attitude, not in others and not in myself. Life is meant to be lived and enjoyed. But for present purposes I am largely setting this perspective to one side. I don't weigh the enjoy-life-while-you-can school of thought against all the supposedly health-enhancing recommendations described in the pages that follow. But this balancing act was always present. I heard it from the experts I interviewed, as well as from friends and colleagues. I think my wife captured the sentiment best when she made the following metaphysically astute observation during the early stages of my diet experiment: "You're a fool. I'd rather eat homemade chocolate chip cookies than be a few pounds lighter." (Her cookies are irresistible, by the way. And as we will see, she eventually changed her mind ... not about my being a fool—that opinion is safe—but about the consumption of cookies.)

Why not engage the philosophically complex cookie conundrum? The optimization of health is widely regarded as an essential public good with both economic and cultural benefits. This is a fact. And whether you agree with the aggressive pursuit of a healthy lifestyle or not, the manipulation of the relevant science remains a bad thing. Generating and distributing misleading information about a topic so central to our world creates inefficiencies and misperceptions, and almost certainly leads to bad choices: by governments about health and research priorities, by health-care providers about treatment options, and by individuals about diet, exercise, and the use of remedies. Being informed is good. Being informed with reliable facts, whether you act on them or not, is very good. My new pub buddies ought to be able to make informed decisions about the actual ramifications of their beer-infused, cigarette-smoking lifestyle. They're free to do whatever they want, but they should still be able to access accurate information.

All of which leads me to my second disclaimer, this one related to what I want you to understand when I use the term *the facts*, or any closely related phrase.

In the pages that follow, I will spill a considerable amount of ink in a critique of the way scientific data is produced and presented. But do not mistake this skeptical analysis for a devaluation of science as a way of understanding the world. On the contrary, I believe that scientific thinking and the use of scientific methods are the most important tools available to us in the exploration of health issues. I reject the idea frequently associated with postmodern philosophers that all knowledge is relative. We can use science to explore why apples fall, airplanes fly, and people get fat. The data that emerges from scientific inquiry (which, at its core, is nothing more than a systematic and verifiable approach to the collection of knowledge) is qualitatively different from assertions that are informed solely by personal experience, instinct, or faith. Yes, it is true that researchers often come to wrong conclusions and that conventional wisdom frequently needs to be revised. But this is, in fact, one of science's greatest strengths. It is self-critical, self-correcting, and eternally evolving.

That said, I respect the postmodernists who are suspicious of those with the hubris to claim privileged access to some immutable truth. And I fully agree with the postmodernist caution that scientific data emerges in a social context that shapes its direction, conclusions, and perceived relevance; that reality is, in fact, one of the themes of this book. Science is rarely pure. There will always be forces—money, pride, ideological leanings—that pull science away from its moorings of dispassion, independence, and objectivity.

Nevertheless there are objective facts about our world that can be revealed. And the use of science, good science, can nudge us closer to seeing and understanding those facts.

And so my final disclaimer is this: I will not go to any great lengths to defend science. Yes, I championed it in the preceding passage. And yes, science, along with M&Ms, denim, and my wife's zucchini bread, ranks among the greatest achievements of the human species. But science does not need me to defend it. In all the topics we are to explore in this book—fitness, diet, genetics, and health remedies—the value of science is largely taken for granted. The language of science is used by those who advocate and market fad diets, exercise routines, and alternative medicines. Even the detoxification dominatrix whose presentation I described above referred to scientific facts throughout her clean-colon sermon. They were false facts, of course. But she turned to scientific-sounding rationales because she knew this would give her pitch credibility.

So even if you don't share my confidence in and reverence for science, all that follows is still relevant. Even if you think we should use the movement of the planets, the reading of tea leaves, or your gut-informed instinct as the primary source of knowledge about health, the twisting of the science about your health is worth understanding and taking into account. Like it or not, most of the information we get about health is projected through the lens of science (a science that is often flawed or forged). And science sits at the centre of all public debates about health, be they about the role of genetics, nutritional guidelines, the value of exercise, or the efficacy of alternative medicines. A clarification of what the best available science tells us helps us all. It allows us to be informed participants in making decisions about both our own health and the things our society should do to facilitate the health of the population—even if, in the end, you'd rather listen to your astrologer. After all, as Mark Twain said, "Get your facts first, then you can distort them as you please."

1

FITNESS
SMARTER! FASTER! STRONGER!

We are constantly told we must be fit. We hear this from doctors, teachers, relatives, governments, employers, and annoying, self-righteous colleagues (i.e., people like me). But what, precisely, *is* fitness? Is it the ability to run a fast marathon even though you look like a sickly, starved waif? Is it the ability to lift 500 pounds despite a BMI that categorizes you as obese? Is it the ability to do the downward-facing dog on a yoga mat? Is it looking good in a thong bikini? Or is it the capacity to have fun chasing your kids around a park? All of these things are a measure of some kind of physical ability. But are they attributes of fitness?

I'll tell you what fitness is: sex. Sex and good abs. Or, more accurately, the cultivation of good abs in order to get lots of good sex. Just writing these sentences makes me want to check my abs (they need work). Go to your corner store. Look at the magazine rack. Staring back at you will be ranks of toothy, tanned, energetic, and fit-looking individuals all showing off their abs.

To test this insight I recently strolled into a smallish airport convenience store and stood in front of the magazine section. Not including the sport-specific magazines, there were 13 fitness-oriented publications: *Health, Best Health, Shape, Abs* (seriously, the magazine is called *Abs*), *Self, Oxygen, Women's Health, Men's*

Health, Fitness, Men's Fitness, Muscle and Fitness, Fat Loss, and *Weight Watchers*. What did I see? A wall of abs, lycra, abs, great teeth, abs, sexiness, and, just to round it off, more abs. All the magazines had cut and beautiful individuals on the cover. Most were doing a seductive torso twist. And all but two of the cover models were displaying their crazily perfect, fat-free abdominal muscles. The two magazines that weren't showing taut tummies were *Health* and *Weight Watchers*. The former didn't count, though. The cover model was the semi-famous actress/singer Zooey Deschanel. She probably didn't want to put her possibly less than ideal belly on display. As for the lack of abs on the cover of the *Weight Watchers* magazine, I suspect that the publishers simply didn't want to scare off their audience.

The sell lines on the covers reinforced the abs-centric imagery: "Flat Abs Instantly!"; "Flat Abs Fast!"; "Flat Sexy Abs!"; "Get Flat Abs as You Shed 9 Pounds!" (why *nine* pounds?); and "8-Pack Abs!"

It has been noted by many academics that fitness in our modern world is largely a commercial enterprise. This is, no doubt, absolutely true. Indeed, the fitness industry has done much to define fitness and what we are supposed to do to be fit. But it's an industry based on sex—or, at least, on the idea that you'll get it if you have the requisite Henckels-knife-sharpening abdominals.

What does this observation have to do with my search for sound health information? Well, good sex is undoubtedly good for you, but selling fitness as sex masks why fitness is so important to health in the first place. And the emphasis on sex and sexiness creates expectations that undermine the health goals associated with exercise. If having sex-producing abs is your fitness goal, you're likely to be disappointed.

Unlike many topics covered in this book, the evidence regarding the benefits of exercise is beyond unequivocal. Exercise is good for you. In fact, if I were forced to pick three courses of action everyone should adopt for the sake of their health they would be: don't smoke, exercise, and do your best to eat well. Everything else is fiddling at the margins. But the big question, and the one I want to explore in this chapter, is: what kind of exercise?

As we will see, the simple truth about what we should do to keep fit is distorted beyond the point of recognition. Many of us are either not doing enough (no surprise) or doing the wrong kind of exercises.

Study after study has demonstrated the benefits of exercise. The Public Health Agency of Canada notes that physical activity appears to reduce the risk of more than 25 chronic conditions, including coronary heart disease, stroke, hypertension, breast cancer, colon cancer, type 2 diabetes, and osteoporosis. And exercise is not just good for the maintenance of normal health and the avoidance of numerous scary ailments. Studies have also shown that exercise during pregnancy is good for both mom and baby. Exercise can confer a clinically meaningful quality-of-life benefit for individuals being treated for various kinds of diseases, including cancer. It *is* good for your sex drive. It has been shown to provide huge psychological and emotional benefits, such as reducing the incidence of stress-related illnesses. In fact, a number of studies have found that regular exercise can be as effective as medication in the management of some psychiatric conditions. Exercise has also been shown to help brain function. One study found that regular exercise actually increased brain volume in the elderly, thus helping stave off the cognitive decline that can accompany aging. Another study, published in the *Proceedings of the National Academy of Sciences*, found that exercise can even boost IQ. Etc., etc., etc.

Exercise is just good for you. It will make you smarter, faster, and stronger. In 2009 England's chief medical officer stated that the benefits of regular physical activity for health, longevity, and well-being "easily surpass the effectiveness of any drugs or other medical treatment." The evidence is simply incontestable.

New research keeps adding to the list of the many ways in which physical activity can benefit your health and well-being. But, again, much of our confusion revolves around what kind of exercise. What should we all be doing to "get fit"? Do we really need sexy abs? If so, how can I get a set?

My search for the answers to these questions began with an exploration of what emerging research tells us is the most effective way to get fit. I uncovered some surprising conclusions that expose the harsh fact that many of us are probably getting it wrong. We will also see that our desire for an easy path to a particular sex-enhancing appearance affords commercial interests the opportunity to twist the truth about fitness. This chapter is, in a nutshell, the saga of the elusive sexy abs.

The search for the truth about fitness took me across the Atlantic and back again. My quest started and finished in cities with a deep commitment to the role of sex, if not exercise, in our lives: Las Vegas and Hollywood.

I got off the elevator moments before the conference was scheduled to start. Surprisingly, and disappointingly, there was no coffee at the event. I was a tad "tired" from the previous night's Vegas activities, and coffee was an absolute must. In my academic career I have probably attended 500 conferences and workshops, and I've never been to one that did not serve coffee in the morning. But then, this was the first time I'd been to a conference where virtually every attendee, not to mention most of the speakers, wore form-fitting lycra. You don't often see that attire at academic conferences.

The event, the National Strength and Conditioning Association's Personal Trainer Conference, attracted trainers and coaches, both novice and well established, from all over the United States. Like me, they came to hear the latest research on exercise and to learn about the fitness industry. This was my first time at a personal trainer conference, and my initial impression matched my expectations—I was among a horde of enthusiastic and fit-looking individuals.

I decided that caffeine was more important than the first ten minutes of the opening talk. I hopped back in the elevator and headed for the main floor, which, this being Vegas, was a casino. The contrast between the bouncy, bright-eyed fitness crowd at the well-lit conference and the smoking, hunched figures sitting in the din of the casino was stark. I travelled from "happy and shiny" to "desperate and sad" in one short elevator ride. There's something uniquely depressing about early-morning gamblers. Had they stayed up all night or did they wake up early? Either way, the prospect was grim.

Another man had slipped into the elevator with me, one of the few in the vicinity not wearing workout gear. He was Todd Miller, a professor in the Department of Exercise Science at George Washington University. Miller comes from a strength-training background (he's an elite-level weightlifting coach), but much of his current academic research is focused on people's attitudes to working out. He was precisely the kind of expert I had hoped to meet at the conference, and, at this moment, we had the same objective. Coffee.

As we made our way to the casino café, dodging a few semi-drunk slot-machine aficionados, Miller and I dove into a discussion about fitness and the fitness industry. I explained my quest, and Miller, who is gregarious, fun, and marvellously opinionated, immediately weighed in with this nugget: "People don't give a shit about health. People don't say 'Wow, you have great blood pressure' or 'Check out that chick's cholesterol.' They may say they care about being healthy, but they really don't give a shit. People want to look good and they equate looks with health. The entire fitness and physical activity industry is built on this reality. It's driven by aesthetics."

It's hard to argue with Miller on these points. Every facet of the industry seems to be either implicitly or explicitly centred on the idea of looking good. And research has shown that people do, in fact, care mostly about appearances; this is one of the biggest motivators for exercising. People exercise not for the physiological benefits, but for weight control and looks. These reasons are particularly dominant for women. Yes, there are vast cultural and age-related complexities associated with the issue of exercise motivation (for example, as we age the health and well-being concerns become more salient), but looks and weight control (for the purpose of looks) are constant themes in almost every study.

A British survey nicely highlights the degree to which appearance, rather than physiological benefit, is a primary driver in this context. The researchers surveyed both men and women about the methods they would choose to lose excess weight. For women, liposuction was the most popular choice (34 percent). This was followed by gastric bypass (26 percent) and cutting calories (14 percent). Only 2 percent of the women surveyed said they would prefer to lose weight by exercising. In other words, eight out of ten women would rather have some form of drastic and possibly dangerous weight-loss surgery than lose weight through diet or exercise. Why? The women said liposuction would "work immediately" and "give great results." But achieving your aesthetic goals via surgery means that your "great result" will not include the innumerable health benefits that come with exercise. You may look good, but you sure won't be fit.

This preoccupation with aesthetics can be found in other health-related behaviours. A 2007 study from Australia found that the desire to "improve physical appearance" was the most influential factor driving food purchases among 18- to 30-year-old women, ranking above factors such as "health benefits" and "low in cholesterol."

There are, of course, abundant examples of people engaging in life-threatening activities simply to achieve a particular look, smoking being the most obvious. Research has shown a strong relationship between body image, a desire to control weight gain, and smoking. For example, a US study followed almost 300 girls for four years and found that those who most wanted to be thin were also most likely to smoke. The girls were using the (hoped-for) appetite-suppressing attribute of smoking as one strategy to achieve their aesthetic goal.

It is a striking paradox that a primary motivation for taking up one of the most damaging activities imaginable, smoking, is the same as the motivation for performing one of the healthiest activities, exercise. We smoke to look good. We exercise to look good. And, if you're Brad Pitt, you do both.

This leads us to the first, and rather cruel, truth about exercise—a truth that is so depressing that I can see why no one is keen to spread the word. My stated goal, however, is to uncover the facts, so here is a crummy dose of reality: it is, in fact, incredibly difficult to drastically change your appearance through exercise alone. Unless you opt for illegal pharmaceutical help (steroids, growth hormones), it requires a massive, long-term commitment to actually change your physique. We are talking Olympian efforts. Flat, sexy abs are not just around the corner. Flat, sexy abs cannot be obtained instantly. Do not believe otherwise. It is a misleading message that leverages our desire for sex (or, at least, sexiness) in order to sell magazines and fitness and diet products. Even if you are blessed with the right genes (which is rare), those abs are probably an Everest-climb of exertion away. And many people may be incapable, due to their physiological allotment, of ever summiting Mount Six-Pack.

In fact, recent research says you cannot use exercise as a principal means of weight loss (and, by the way, slow, steady weight loss is the only way, the absolute *only* way, to expose your gut muscles in a manner that will give you the classic six pack). You will note that my list of the benefits of exercise has no mention of weight loss. Yes, you might read stories of big weight-loss successes; many can be found in health magazines, on blogs, and on TV. But, in general and in the long term, the data simply does not support the use of exercise as a primary tool for getting thin.

Todd Miller ranks the idea of weight loss as one of the biggest myths associated with physical activity. "People don't understand that it's very difficult to exercise enough to lose weight. If that's why you're doing it, you're going to fail. In part, it's because you're fighting creeping obesity. Everyone puts on weight as they age. If you're keeping your weight constant, you're winning the battle. Working out is actually working."

It's not a very sexy message, is it? Exercise hard so you can … *stay the same!* Can you imagine this headline on a fitness magazine's cover: "Do This Gut-Busting Workout and Slightly Increase Your Chance of Not Putting on Weight!" Or: "Spend a Mere One Hour a Day Exercising Pretty Darn Hard and You Too Can Look Kinda the Same Next Year!" Yippee. But, sadly, the evidence suggests that these wan promises are more truthful than most that you'll see.

My colleague Kim Raine is a marathon runner. A good one. Her best time is just over three and a half hours. She is also a public health and nutrition expert. One day we were talking about the recent research on the myth regarding exercise and weight loss and she declared that this has been a source of personal frustration. "I've run 18 marathons and I put one pound *on* for each one. Eighteen marathons and 18 pounds heavier. It is so maddening!"

How can this be? Why would such a fit woman put on weight while training for marathons? Why, speaking more broadly, is it so difficult to exercise enough to lose weight or change our appearance?

First and foremost, humans are incredibly efficient eating machines. We don't need many calories to do extremely difficult tasks. If you want to lose weight, you must burn more calories than you take in. Period. That is the only magic formula. And while exercise burns calories, it does not burn as many as people think. If you start your day with a Starbucks latte (270 calories) and one of their Classic Blueberry muffins (470 calories), you will need to run for well over an hour at a pretty intense pace (not jogging) just to break even. For the rest of the day you must eat less than your body needs or you won't lose an ounce. If you eat a banana and drink 20 ounces of a sports drink after a 45-minute tennis game, you are breaking even—you haven't achieved a single weight-loss benefit. Zero.

And, by the way, you should probably ignore the calorie counts you see on the exercise machines. They are likely wildly generous overestimations of what you are actually burning. If the treadmill, stationary bike, or elliptical machine says you have burned 400 calories, a safer bet would be 200 or 300. I am not saying the evidence is definitive regarding all machines (though many commentators seem to agree); I simply don't trust them. Naturally, there is a marketing incentive for conveying the perception that the machine burns lots. Another subtle twist in our midst.

On occasion, I enjoy a big bag of M&Ms when I go to the movies. Okay, I devour a bag at absolutely *every* movie I go to. I exercise a lot, but that bag of candy-covered peanuts—about 1200 chocolatey calories—probably neutralizes the weight-loss benefits of two or three workouts. If I go to two movies in a week (not uncommon), I am calorically sunk. Unfortunately, I love movies. And I love my M&Ms. It's even possible that I love movies *because* I love M&Ms. A Pavlovian cinematic experience.

So when you hear someone say "I work out so that I can eat what I want," you should know that they are deluded, unless: (a) they are training as hard as a Tour de France cyclist but don't care about their weight (a highly unlikely combination); or (b) "I can eat what I want" means the unbridled consumption of broccoli, celery, water, or air.

The second, and slightly more controversial, reason that exercise alone will not lead to weight loss is that working out increases appetite. And because it makes you hungry, it is tough to keep your daily calorie intake in the negative zone. As crazily efficient consumption entities, we humans are biologically calibrated to stay at whatever weight we are at. Our bodies crave the status quo (even if the status quo is kind of doughy), a reality confirmed by both animal studies and clinical nutrition research. If we burn off calories on a long run, our bodies tell us to eat more to make up the difference.

I say that the idea that "exercise makes us eat" is somewhat controversial principally because the data is not completely settled. Some studies suggest that intense exercise can, in fact, suppress appetite or extend the satiating effects of eating. But this effect is not certain, or, at best, it is not dramatic, and, if it does exist, it seems likely to be overwhelmed by the belief that we can eat more when we exercise. And, sadly, the hunger-inducing effect of exercise *does* seem to exist for women. A 2009 meta-analysis (that is, a study that rigorously looks at all the available data) on this issue concluded that acute exercise has no effect on subsequent caloric intake in men (it did not increase or decrease appetite), whereas "in women an increase in energy intake [i.e., calories] is usually observed, either decreasing or abolishing the effects of exercise" on overall calorie burn. Given the social pressure on women to be thin, this seems a cruel evolutionary twist.

A study published in the *Journal of the American Medical Association* highlights the modest role of exercise in weight loss and weight management. This impressive study followed more than 34,000 women for 13 years. It found that the women who did at least one hour of moderate-intensity exercise every day were the individuals most likely not to put on weight. Most *still* put on weight, mind you. But those who did an hour of daily exercise were the most successful at maintaining their weight. The media reaction to this research was fascinating and telling. The study was, as first, presented as good news. For example, one newspaper account published on the same day that the study was released optimistically declared: "Being Active an Hour a Day Puts Brakes on Weight Gain." But very soon the reality of the study's underwhelming message kicked in. "Hour of Exercise? Yeah, Right, Many Women Say" was a headline published days later.

This study also highlights the point Todd Miller made about "obesity creep." As we get older, we put on weight. This inevitability has been known for a long time. For example, a study done in the late 1990s followed approximately 5000 male runners. These were active, physically fit running nerds. You know the type. But even within this group of health-conscious individuals there was a depressingly steady pattern of weight gain. The study showed that in runners between the ages of 18 and 50 weight gain occurred at the same rate almost regardless of the number of miles run per week. Per decade, it worked out to about 3.3 pounds and about 3/4 of an inch around the waist.

So, if you are exercising and you are not putting on weight as you age—or only putting on a bit—you are actually doing very well compared to most. This reality is both interesting (why doesn't everyone know this?) and depressing in that receding-hairline kind of way. It's also sobering in that society-is-becoming-irrevocably-obese kind of way.

Back in Las Vegas, with our jumbo, casino-priced coffees in hand, Todd and I headed back up to the conference. There were a variety of presentations going on at the same time. Some were business oriented (what works to make money off your clients). Some focused on motivational training advice (what makes your clients work). And some provided new data on fitness (what actually works). I opted for the latter and decided to take in a talk by Mark Peterson, a young researcher from the University of Michigan.

Though Peterson is built like a personal trainer, he talks like a professor. His presentation was heavily data driven. Charts, graphs, and equations. At first I found it a bit hard to follow, but I also sensed he was providing the kind of evidence I was after. He studies the best way to get benefits from exercise. And while his research is aimed primarily at the elderly, his results have relevance for us all. Indeed, his conclusions are surprising and refreshing. They cut against what so many of us think of as the kind of exercise that makes us fit.

Before I get to Peterson's conclusions, however—conclusions that build on a growing body of evidence—I need to define "fitness." This is a concept that has been severely distorted by the media and by commercial interests. I am searching for fitness data with health implications, so my definition of fitness does not include flat, sexy abs as the primary end point. For me, fitness includes strength, aerobic capacity, and an improvement in the biomedical markers of health (low blood pressure, good cholesterol, and appropriate levels of blood glucose). It also includes the physiological improvements that ward off disease (such as cancer), even if we don't fully understand the biological mechanism at play. It includes the achievement of a physical state that allows us to simply feel better. In other words, my definition of fitness isn't based on an aesthetic goal, but rather on a physical and, to some degree, mental state. I think it's fair to say that people want to look fit because this look is associated with physiological characteristics that are desirable, but, somewhat perversely, they don't really care much about those characteristics. They just want to *look* healthy, because to look healthy is to look desirable. For my purposes, however, fitness is defined as an actual physical state that enhances one's health, not just the *appearance* of a physical state that enhances health.

For the sake of clarity, here is a practical application of my definition: you can get sexy abs through liposuction, heroin addiction, and agreeing to a regime of near-starvation on a reality TV show, among other radical methods. Obtaining this look through these methods does not make you fit.

The fact that we find certain characteristics sexy, such as flat abs, is no accident. The look has evolutionary underpinnings. Research has shown that men find women with flat abs and a waist-to-hip ratio of 70–80 percent to be the most desirable. This combination sends a signal to the world: the woman with these attributes is young, healthy, and ready to produce babies. To a savannah-dwelling early man looking for a hot date, I'm guessing that flat abs were a particularly important female characteristic because a taut tummy advertised that there was no baby on the way or a new one at home consuming valuable, hard to find, savannah resources. Of course, over the centuries, what we find desirable has been shaped by social norms and fads that are not necessarily tethered to evolutionary needs (although a clear evolutionary advantage was obtained by those of us who wore acid-wash jeans and mullets—I snagged a pretty amazing partner). Despite such cultural tinkering, however, we are all still striving for a body that is reproductively desirable, even if our reproductive years are behind us or, more paradoxically, if we aren't even interested in the end product of the reproductive process. The part of our brain that is tickled by this look does not care, at least on first impressions, how this look is obtained. Of course, a first-date conversation about the wonders of plastic surgery or steroid use may, for some, alter the impact of that first impression.

Luckily, the activities that lead to my defined state of fitness are also the most efficient way for individuals to move closer to that (likely unattainable) evolutionarily shaped and industry-marketed aesthetic goal. If, despite the bleak evidence about the impact of exercise on weight loss and physique-changing muscle development (more on this further on), looks remain your primary motivation for getting active, everything that follows is still relevant. If you want those abs, the exercises that are most efficient at getting you fit are also the ones that are most likely to nudge you a wee bit closer to that ideal.

We have established that you won't lose weight just by exercising. Let's dispense with a few more myths. I asked every fitness expert I interviewed to name his or her favourite misconceptions, and there ended up being so many it would be cumbersome to list them all. Many of these myths are widely known to be untrue, but they nevertheless persist, in large part because they help to sell fitness products. But it is worth highlighting a few of the most common myths, as they are relevant to what we should (and shouldn't) be doing to get fit. The first one—and the one noted by absolutely everyone—is the myth we have already covered, that you can lose weight through exercise alone. As one US exercise expert, Eric Ravussin, was quoted as saying in the *New York Times*, "In general, exercise by itself is pretty useless for weight loss." You need to do more than simply exercise. And that means diet, the subject of the next chapter.

The next big myth is closely related to the weight-loss issue. How many times have you read or heard someone say muscle burns more calories than fat and, therefore, if you train, you can eat more? Don't worry, the thinking goes, your muscles will burn up the calories! Nonsense. There is simply no solid evidence to support this claim. You may burn a few extra calories hauling around your big muscles, but only a very few. Do the math. A pound of muscle burns around six calories as it hangs out in your body during an average day. If you are training hard and concentrating primarily on muscle development (in other words, avoiding the aerobic activities that can make it more difficult to put on muscle), then a realistic goal for muscle gain would be about one to two pounds of muscle per month. This is under optimum conditions: perfect nutrition, lots of sleep, and the good fortune (at least from a muscle-building perspective) to be a young male. A more realistic goal for those of us who can't spend hours in the gym would be less. And, of course, as we get older and the hormones that help us build muscle, such as testosterone, decrease, it becomes more difficult to pack on the muscle-infused pounds. But let's say that by sheer determination, or with pharmaceutical assistance, you managed to put on ten pounds of pure muscle (which, for most healthy mortals, would likely take more than a year of very hard work): you could then consume approximately 60 additional calories a day. In cookie terminology, that's one Oreo. Live it up. Don't get me wrong, having more muscle is a good thing and resistance training does burn calories and increase your metabolism, but don't think it will allow you to eat whatever you want.

A side observation to this muscle myth: people often tell me they have put on weight working out and quickly follow this with the "muscle weighs more than fat" explanation. This is almost certainly total baloney (or fatty food substance of your choice). Given how difficult it is to put on large quantities of muscle, a much more likely explanation is that the exercisers are ingesting more food because they think they can. And the additional calories become more fat. If you are working out and over the course of a few months you put on weight, it is almost certainly fat. I myself have fallen prey to this myth. In 2004 I got very serious about track cycling. I'm talking family-neglect serious. I was training with an Olympic gold medalist and her coach. I was focusing on a sprint event, so the routine involved two-a-day training sessions and vast amounts of weightlifting. Over the course of a year I put on well over 20 pounds. I assumed it was solid muscle. I ate more. I ate protein powder. I ate many big meals. I chugged sports drinks. Every single one of these behaviours was unnecessary and blubber-building. Despite my incredibly intense training, my percentage of body fat increased substantially.

Another myth that every expert railed against was the idea of "spot reduction." Both academic researchers and renowned personal trainers are enraged by this delusion. It is one of the most pervasive myths and the one tied to the most fitness products. Here's the straight goods: you cannot lose fat in a particular region of the body by working that part of the body. You cannot "tone." Wearing special shoes or doing particular exercises will not *tone.* You cannot lose stomach fat by doing sit-ups. You cannot get Michelle Obama's arms by working your triceps. You cannot burn butt fat by doing leg lifts. The Ab-Flex-Spinner-Thingy will not tighten your tummy. The *only* way to get the *look* of toned muscles in a particular region of the body is to lose enough fat to expose the shape of the muscles in that region. And the *only* way to do that is to lose weight. And here's more harsh truth: you must possess a very low body fat percentage (around 10–12 percent) to actually get that "toned" look we are all after. To be fair, you also need to *have* muscles in the region of choice, but that really is a secondary concern. There is an iconic photo of punk rocker Iggy Pop surfing the crowd at Chicago's Soldier Field in 1970. He is so perfectly toned that he could be a cover model for one of today's fitness magazines. But I guarantee Iggy acquired this frame via genetics and partying-induced malnutrition. It wasn't weight training, a healthy diet, and bi-weekly meetings with a personal trainer. Sid Vicious, another rocker who lived on the edge, was pretty cut too.

The next big myth was new to me. How many hours have you spent stretching? Doing toe touches, side splits, and calf stretches? Add it up. For me, it is a lot. If I pick my most sports-involved decade, from ages 12 to 22, I probably stretched six days a week. To account for holidays and lazy spells, let's say it was four days a week for 15 minutes a day (a very conservative estimate). That's an hour a week for ten years. Over 500 hours of doing something I hate. We did it in elementary school gym class. We did it as part of the warm-up for almost every organized sport. You see pro athletes do it on the sidelines. And for some fitness activities, it is a central part of the regime. But, as it turns out, stretching is not only largely useless, it might even be harmful. Ryan Rhodes, a behavioural medicine professor at the University of Victoria, says the emphasis on flexibility is a major fitness misconception. "People spend so much time on this. But there is no evidence regarding stretching benefits. It doesn't help injuries. It doesn't help performance. You just need basic range of motion."

Comments such as Rhodes's, which I heard from many I talked to, motivated me to look into the stretching literature (yes, there *is* a stretching literature). And the skepticism about the value of stretching is justified. There is surprisingly little data to support spending time touching your toes. For example, numerous studies have shown that stretching does not help, at least not in any significant way, in the prevention of injuries (which is, of course, the reason it is recommended). Different studies have featured conclusions such as "The combined risk reduction of 5 percent indicates that the stretching protocols used in these studies do not meaningfully reduce lower extremity injury risk" and "Strong evidence exists that stretching has no beneficial effect on injury prevention in these sports." A 2004 review of over 350 studies done by the US Centers for Disease Control and Prevention found that "stretching was not significantly associated with a reduction in total injuries." There is even some evidence that stretching can hurt athletic performance. For example, a 2008 study of elite track athletes found that stretching inhibits sprint performance.

Numerous studies are starting to pop up that suggest that stretching is bad for a range of athletic activities. If you made a list of activities associated with flexibility, rhythmic gymnastics would need to be near the top. Yet studies show that deep stretching before a competition is bad even for this cohort of athletes. One 2010 study found that those who stretch before a competition got lower scores, likely because stretching hurt the ability to leap.

To be clear, the data is not conclusive on this issue. There just isn't enough evidence to get excited about the activity. Some studies do support the practice, though the endorsement is usually fairly tepid. For example, a 2009 study of almost 2400 active adults found that "stretching before and after physical activity does not appreciably reduce all injury risk, but probably reduces the risk of some injuries, and does reduce the risk of bothersome soreness." This is hardly a ringing endorsement for an activity that has been an exercise staple for decades. Indeed, given the strength of the conventional wisdom—that stretching prevents injury—the data is amazingly equivocal. What does this mean for you and me? From the perspective of health and fitness, a big emphasis on stretching does not make sense. You can likely get the flexibility you need in your daily life or in the sports you are involved in by simply doing exercises that provide a nice range of motion, such as lunges and squats. At least a few studies have found that those kinds of exercises are *better* at producing and maintaining flexibility than more traditional methods.

So, kill the toe touch. If you enjoy stretching or you are an acrobat with Cirque du Soleil (or want to date someone from Cirque du Soleil) and must be able to stick your foot in your mouth, then by all means, stretch up a storm. But if you are pressed for time and simply want to get fit, stretching should not be high on the priority list.

So what *should* be high on the priority list? This question brings us back to Mark Peterson's presentation in Vegas.

During his talk he blasted through some 75 PowerPoint slides. I listened while standing in the doorway of the lecture hall, mostly so that I could gaze out at the Strip during the boring bits. This turned out to be a mistake. It wasn't boring, and there was so much I wanted to remember that I found myself struggling, while standing, to write them down. (My notes remain almost unreadable.) Mark was speaking quickly and I could tell that more than a few members of his audience—a mix of personal trainers, students, and academics—were a bit confused. But as he proceeded through the data, both from his own work and that of other academics, the message became clear. If health and fitness are your goals, you should be lifting heavy weights at least a few times a week. Not soup cans or tiny pink barbells. Real weights. Challenging weights. Big-time iron.

As soon as Mark finished his talk I grabbed him for a follow-up chat. We talked about the value of all the things people do to get fit: running (over-emphasized), stretching (a waste of valuable exercise time), and toning (a fantasy the mere mention of which caused Mark to roll his eyes, shake his head, and declare: "There is no such thing as a toning exercise!"). But mostly Mark reiterated the main message from his talk: it's all about doing challenging resistance training. He is suspicious of the value of low-intensity weight training. "It is only good for getting people started. The true benefits come from working hard and with intensity. And this can and should be done at all ages."

To emphasize his point, Mark told a remarkable story. For one of his studies he worked with an 83-year-old woman. When she first came to see him she had trouble getting out of a chair. Mark got her working with weights. Soon she was doing power lifting (the kind of weightlifting you see at the Olympics). She can now squat 135 pounds, more than her own weight. She won gold at the senior Olympics in power lifting. (Okay, it turns out she was the only one in her age class, but it's still pretty impressive.) This woman has more energy, is more mobile and more confident, and now has a closet full of those puffy sweatpants with cartoon characters on them that most weightlifters seem to wear. Actually, the bit about the sweatpants isn't true (as far as I know), but there seems little doubt that weightlifting vastly improved her quality of life.

I've known for a long time that resistance training is a great way to gain strength. But Mark's presentation highlighted the degree to which intense resistance training should be a central part of almost everyone's fitness routine—not just jocks and people who want to look good naked.

When we think of serious resistance training, what often pops to mind is a buff Arnold Schwarzenegger and, for the current generation, the meatheads in the TV show *Jersey Shore*. We have been trained to think of great fitness activities as primarily aerobic: long runs and bike rides. Canada's ParticipACTION website, for example, a government-sponsored physical fitness program, states: "We all know great ways to get physical—jogging, rollerblading, hiking, skating, skiing and so on." I asked friends and acquaintances what activities the term "fitness" brought to mind. I found a remarkably consistent response: an endorsement of the long, slow aerobic stuff and a tremendous resistance to resistance training, especially from women and those involved in endurance sports. Mark has noted this too. "Aerobics exercise is always pushed as most important. There is a perception that strength training is for big weightlifting guys. This is wrong. In fact, women and the elderly are the ones that benefit most from resistance training, not young healthy men."

Weightlifting provides a wide range of health benefits it's not usually associated with, such as lowering cholesterol and blood pressure. In a 2006 article in the *Canadian Medical Association Journal* that reviewed all the evidence of the benefits of exercise, the authors characterized the research on the value of resistance training as supporting a "paradigm shift." They noted that studies have "revealed that people with high levels of muscular strength have fewer functional limitations and lower incidences of chronic diseases such as diabetes, stroke, arthritis, coronary artery disease and pulmonary disorders ... Musculoskeletal fitness is positively associated with functional independence, mobility, glucose homeostasis, bone health, psychological well-being and overall quality of life and is negatively associated with the risk of falls, illness and premature death." A greater emphasis on this aspect of fitness is highly justified, the authors concluded.

One 2009 study from India found that resistance training "is a more effective form of exercise training than AE [aerobic exercise] for improving glycemic control, blood pressure and heart rate in type 2 diabetics." Also, if done with intensity, resistance training produces cardiovascular benefits similar to typical aerobic activities. A small 2010 study out of the United States found that intense resistance training actually produced greater aerobic benefits, such as reduced blood pressure, than traditional aerobic training. There are even studies that support the use of resistance training as a means of rehab from heart surgery and other traumatic events. Other studies have shown it to be as effective as aerobic work for weight management (which, as discussed earlier, is not necessarily an impressive statistic, but still a step in the right direction). Contrary to conventional wisdom, even serious endurance athletes, such as cyclists and distance runners, can benefit from challenging resistance work. A 2010 study of highly trained cyclists found that replacing a portion of endurance training with resistance training resulted in improved time trial performance and maximal power. (Note to endurance athletes: the key is to devote a day to lifting. You must truly *replace* the endurance work with weights. You shouldn't do a mix.)

Surprisingly, resistance training also appears to be good for kids. When I was growing up, the conventional wisdom was that pre-adolescent children should stay away from weights. It was widely believed that resistance training would hurt young bones and joints. Untrue, according to recent research. A meta-analysis of relevant evidence, published in 2010 in the journal *Pediatrics,* found that resistance training is beneficial for both adolescents and children. Indeed, some experts believe that given the sedentary lifestyle of many of today's children, resistance training should be viewed as essential. It helps to build strength that will allow kids to be more active and avoid injury.

I am not saying that individuals should avoid aerobic work or that it is not beneficial. On the contrary: aerobic exercise is essential. But if I were forced to pick one activity to place high on the priority list, perhaps even at number one, it would be intense resistance training. You can get almost all the health benefits associated with fitness from resistance training. This is not true of aerobic workouts. You can do a resistance program in a relatively short amount of time, and, if you do lifting movements that put you through a full range of motion, you get all the flexibility you are likely to need for the activities of daily life. For the elderly (and for those of us moving slowly toward that demographic), it helps with a range of quality-of-life issues, such as mobility and injury prevention. Research, including work published by Mark Peterson in 2011, has shown that strength can be gained at any age and that it is one of the single most important factors in the maintenance of function and, thus, a high quality of life. Plus, resistance training, if done properly, is not as associated with the overuse injuries so common to activities like running. A 2010 review of the benefits of exercise done by the Swedish National Institute of Public Health concluded that "the collective assessment is that strength training is at least as safe as aerobic training if not safer." Indeed, research has shown—again, contrary to conventional wisdom—that lifting can improve the health of your back, even if you are showing signs of spinal degeneration. This point was recently confirmed in an award-winning 2010 study from the University of Alberta. The authors found that by challenging the spine with weights you will help to keep it healthy. As the lead author told me in an email, regarding the spine, "Use it or lose it."

So, what do we mean by resistance training? While lots of activities are healthy and worthwhile, there are a few that maximize the health benefits and inch you closer to the ownership of sexy abs. The first rule is to follow Mark's advice about effort. Work hard. What does this mean? You should lift weights heavy enough to make 8 to 12 repetitions difficult. The resistance should be pretty significant. Doing a million repeats of a light weight is not the best strategy. One 2009 study from the University of Michigan concluded that many of us are not pushing ourselves hard enough. The researchers followed a group of gym-goers and found that most of them selected weights that were too light to stimulate muscle growth. Virtually all fitness improvement—strength, cardio, and even flexibility—comes from a process called adaptation. It's easy to understand, and it makes perfect sense: you must challenge your body enough to stimulate a biological change. For weightlifting, this means you must break down your muscles so that they regrow stronger.

What else is involved in a resistance training program? The second rule is to do activities that engage more than one part of your body. It does not make sense to stand in a gym and work only your biceps. Do an exercise that works all of your body, especially the big muscle groups. For example, a chin-up (or some kind of rowing or pull-down exercise) works the big back muscles, the core, *and* that all-important beach muscle, the biceps. Also, try to do exercises that put your body through a range of motion. This will give you the strength and mobility for a wide variety of sports and the activities of daily life (such as, in my daily life, grabbing a screaming kid for a time-out or lifting an airport toilet seat with your foot). In fact, you can build a killer fitness routine by doing variations on the following four basic exercises: the squat (change it up with lunges); the bench press (or push-ups); the chin-up (or some kind of pull-down or rowing move); and an exercise that has been called the king of all exercises because it challenges almost every muscle in the body, the deadlift (this is, basically, lifting a heavy weight off the floor). While you must be careful to do these exercises properly, all are easy to learn (it's easy to find instructional videos on YouTube, for example) and you don't really need fancy equipment, especially when you're just getting started.

Third, in order to maximize both the fitness benefits and time efficiency, do your resistance training as a circuit. In other words, move from one exercise to the next without resting. For example, do a squat move and then go straight to a rowing or pull-down exercise. This will keep your heart rate high, thus giving you more of an aerobic boost. It's also efficient. As noted in a 2010 study that compared circuit training with other methods, "Incorporating this [circuit] method of resistance exercise may benefit exercisers attempting to increase energy expenditure and have a fixed exercise volume with limited exercise time available." In other words, if you're busy, your best fitness bang for the buck is circuit training.

Finally, you should mix it up after a few weeks. You need to constantly challenge and surprise your body to force it to adapt. So, increase the resistance or change exercises slightly. Don't do the same thing over and over again. If you are becoming comfortable with a particular routine, this is an indication that you should change something. If your body has adapted, then force it to adapt all over again to something new.

That's it. Simple. Do intense resistance training using movements that engage your whole body. If overall fitness is the goal, we should all be doing this kind of thing, or some type of physiological analogue, at least a few times a week.

Of course, there is more to fitness than doing some resistance-based circuit training. While some (but not all) of the experts I talked to thought aerobic activity was often overemphasized, everyone agreed it was essential, the foundation of good fitness.

To find answers regarding the best, most efficient way to get aerobic fitness I decided to leave Las Vegas (where aerobic fitness is mostly about maintaining the stamina to sit at a slot machine for 12 hours). But as I was leaving, I made what I'd call a reality-check observation. There were dozens of personal trainers at the conference, many of them among the best in their field. These were individuals who make their living in the fitness industry. Not only did they have a realistic understanding of what is required in order to get fit, they were also highly motivated to *be* fit—a personal trainer who looks out of shape is hardly a good advertisement for his or her services. And yet, despite this occupational reality, there were clearly just a few who would pass the "sexy abs" test. Yes, there were a number of young trainers with ridiculously cut stomachs. But the reality was that most of the conference participants simply looked like healthy versions of normal. Some were slightly softer than ideal, some were rail thin, and some could have passed for bank tellers. As I left the conference hall I actually wondered, If we lived in a universe where sexy abs really could be obtained instantly or even with a bit of effort, then why didn't everyone at this event possess them?

Exeter, England, might very well be the anti-Vegas. I am willing to bet no one has ever called it glamorous. Nor does the word "exciting" come to mind. But it does have charm. Not quite quaint, it feels like a real English city: a bit of grit, lots of history, and some beautiful buildings and countryside. It is also home to Exeter University, an institution that employs several researchers key to my search.

Gary O'Donovan is a lecturer in exercise physiology and program director for the M.Sc. in sport and exercise medicine. He has done research and commentary on a wide range of activities associated with exercise. He also has such a passion for this subject that he was fun to interview and a fountain of great ideas.

I met O'Donovan in a tiny café a few blocks from his office. We chatted about fitness as we each demolished a piece of delicious carrot cake. We calculated that each piece was in the 400-calorie range, an irony hardly lost on us. "That little piece of cake," O'Donovan said, pointing at his empty plate, "is probably going to be about an hour or more of brisk walking. Oh well. It was good."

O'Donovan was instrumental in drawing up *The ABC of Physical Activity for Health: A Consensus Statement from the British Association of Sport and Exercise Sciences,* which was published a few weeks before we met. Co-authored by more than a dozen academics from across the UK, the consensus paper drew together the evidence associated with fitness and health, and laid it out in a series of recommendations aimed at health-care professionals.

And what was the biggest message from the consensus statement? Once again, and in a word, *intensity.* O'Donovan noted that for a long time the message to the public has been that "moderate exercise is all you need for health." But O'Donovan thinks this is seriously misleading. "In reality, the more you put in, the more you get out. Moderate is good, but vigorous is better. There is a clear dose response."

O'Donovan gets agitated by the soft-sell message, which he says often comes from governments, public health officials, and the media. It has created the belief that you can get all the benefits of exercise from a leisurely stroll or a round of golf. Yes, any kind of physical activity is better than none at all. But if maximizing one's health benefit is the goal, "moderation" needs to be replaced by "intensity."

In 2007, O'Donovan conducted an interesting study in which he asked more than 1000 British men and women what was better for them, moderate or vigorous exercise. Fifty-six percent of the men and 71 percent of the women thought that moderate exercise offered more health benefits. His study concludes with the bold declaration that "policymakers have an obligation to equip the public to make fully informed decisions about physical activity and health. [Currently the public] erroneously believes that moderate activity offers greater health benefits than vigorous activity."

I asked O'Donovan why the "moderate is best" belief is pervasive. Why do we all think moderate is good enough? Where is this particular misleading message coming from? He believes that public health officials set the bar low so as not to scare people. "On a population level you get the biggest health benefits from getting completely sedentary individuals moving. Getting sedentary individuals to do even a bit of physical exercise has a huge health impact—on the health of the broader population, but not on individuals. So, if you're making policy for a country, this is the key. But it's only half the story. At the individual level, you want to get maximum benefit, and that requires vigorous work. If you do exercise right, you can reduce individual health risks significantly, by as much as 50 percent." In other words, the social marketing of *moderate* has overwhelmed the reality of *vigorous*. This is a twist that occurs in order to sell a public health message.[1]

During the Vancouver Winter Olympics my family and I watched almost every hour of the TV coverage. We saw the Canadian Body Break fitness advertisements so many times that we could repeat them by heart. These advertisements were intended to promote a healthy lifestyle. Two disturbingly wholesome hosts were shown doing something fun, casual, and "moderate." Riding a bike by a lake. Walking in the woods. Playing in a pool. These are all healthy and fun activities. (I'm guessing they're fun. Our family was too busy sitting in front of the TV to do any of this family-oriented stuff.) What all these activities were not was vigorous. I have seen three decades of the Canadian government's ParticipACTION physical activity public messages and I can't recall "vigorous" ever being a big theme. To refresh my memory I went to the ParticipACTION website and viewed all the old commercials. In fairness, the claymation "just do it" advertisement from the 1980s shows some smiling clay figures that look as though they are working pretty darn hard (I think I see some clay sweat). But the more recent animated personal fitness stories, which are beautifully done, emphasize the moderate: ballroom dancing, taking the stairs, walking the dog. One of the stories made reference to a spin class (which I can only assume is vigorous) and another has a mom playing soccer. But, in general, the overall vibe is that moderate is fine, and the take-away message is clear: listen up, you nation of sloths, please, please just do something. Anything.

Another phenomenon O'Donovan and many others have discovered is that even when people *think* they are exercising vigorously, or even moderately, they often aren't. There is a big disconnect between perception and the actual intensity and volume of work. A few years ago the US Centers for Disease Control and Prevention (CDC) conducted a survey in which they asked people how often and how hard they exercised. Forty-nine percent of respondents believed they were exercising regularly at a moderate level of intensity, and 27.5 percent thought they were working out at a vigorous level. Subsequent research using tools that mapped actual activity found that, in fact, only 21 percent of the population was exercising at moderate intensity and only 1 percent was exercising vigorously. (Here is a quick test of vigorous: it should be difficult to talk while doing the exercise—you should just be able to get out two to three words.) A study published in 2010 found equally stark numbers—only 5 percent of participants were conducting an activity that was genuinely vigorous. And what did this study find to be the most common moderate physical activity? Meal preparation!

And so what does this mean for those of us trying to get fit? Well, first, ignore the "moderate is sufficient" misperception. The most efficient way to get the health benefits associated with aerobic activity is to do at least some vigorous exercise. We have to work hard enough so that it is tough to talk. And we should be doing this a couple of times a week. I am not saying that walking or going on an easy bike ride with the family is not good for you. It *is*. Any activity is great. Not to mention that these activities are worthwhile for a variety of different, non-fitness-oriented reasons. But the point for our purposes here is that moderate exercise is not the most efficient way to get the benefits out of physical activity.[2]

Even more specific advice is available. Recent research has shown that a particular kind of aerobic activity is, in the big picture, the most efficient. High-intensity interval training is vastly more effective than the usual steady-state aerobic training. When people do aerobic exercise they often simply run, bike, or swim for a set period of time. If you're a jogger, for example, you may simply put on a pair of running shoes and head out the door. After a short warm-up period, you run at a relatively steady pace, perhaps "kicking" the final mile. Turns out, this is not the best way to exercise; in fact, it can be a waste of time. If efficiency and effectiveness are a priority, you should be doing short blasts of intense activity followed by periods of recovery.

One study found that several 30-second-long sprints on an exercise bike, with a few minutes of recovery between each sprint, three days a week, was as effective as a 40- to 60-minute run done five times a week. This is an amazing conclusion, when you think about it: the interval training resulted in comparable physiological results (they measured stuff like mitochondrial biogenesis ... whatever that is) and it took a fraction of the time. Another study, conducted at McMaster University and published in 2010, found that ten one-minute sprints on a standard stationary bike with about one minute of rest in between, three times a week over two weeks, works as well as many hours of moderate-intensity, long-term biking. From a time commitment perspective, interval training is likely four to five times as effective as traditional endurance training.

Given that "I have no time" is a common excuse for avoiding exercise (in some studies that explored barriers to physical activity, it was found to be the primary reason), this kind of research should be viewed as wonderful news. It requires intense effort—the intervals need to be genuinely hard—but it is vastly more efficient than a long, slow slog.

This is the kind of research that demonstrates how science can inform and improve approaches to exercise. Indeed, in March of 2010, one exercise researcher, Jan Helgerud from the Norwegian University of Science and Technology, said: "This is like finding a new pill that works twice as well ... we should immediately throw out the old way of exercising."

What's more, the interval approach should not be reserved for young, healthy athletes. A 2007 study found that intense interval training was a more effective strategy than moderate continuous training for "improving aerobic capacity, endothelial function, and quality of life in patients with postinfarction heart failure."

So, also high on my exercise priority list: intense intervals. And Gary O'Donovan agrees. In fact, O'Donovan thinks that if you are having a busy week and you can only get in one short workout, he'd vote for "doing an intense interval workout, like running hills or sprints on a bike." For him, "Intensity is the single most important variable. Get out there and work hard."

Intensity is vital to so many of the things we do, but so is discipline, which is precisely why O'Donovan and I congratulated ourselves for leaving some of the carrot cake icing on our plates when we paid up and left the café. Willpower is a key to staying healthy, and it took a lot of it not to lick that plate clean.

Before we examine more of the forces that twist the simple truths about fitness, I want to describe the basics of a good weekly routine. In fact, this is my routine. It draws together all the recommendations from the guidelines and research I collected and the advice I was given. Though there was not complete agreement on everything—O'Donovan, for example, thought intense intervals were the most important element, while others, such as Miller and Peterson, stressed resistance training—I feel confident that all would agree that the following program hits the right notes. I'm placing it here purely for illustrative purposes.

The goal of this routine is to obtain and maintain the kind of fitness that will give you continuing health benefits. It's also structured to make the best use of your time. You may love golf, lawn bowling, softball, long slow jogs, and yoga. You are being active and probably not eating too much while doing these "sports" (although calorie consumption is a pretty common by-product of some of these activities). But, no matter how good these activities make you feel, they are not a great way to get fit. Here are the basics of an exercise routine that is.

Day 1: Short intervals. Warm up for 10 minutes. This should be done before every workout and should include two elements: first, go through a range of movements, such as deep knee bends and lunges, that get the body ready to work and put it through a range of motions; and second, put in enough of an effort so that you begin to sweat. Do not bother stretching. It's a waste of time and probably bad for the workout performance. Then start your intervals. An interval is basically a burst of intense effort. So, go hard (really hard) for 10 to 20 seconds and then rest for 50 to 40 seconds. Repeat four to ten times depending on your fitness level. As you become more comfortable with the routine, do this set two to three times. All intervals can be done on a bike, treadmill, or elliptical machine. They can be done outside while running or biking. If you are just getting started, and don't feel terribly fit, you can do intervals by simply walking and jogging. But remember, work hard.[3]

Day 2: Resistance training. This should engage your entire body. You should push stuff (e.g., squats, lunges, bench presses, push-ups). You should pull stuff (e.g., lat machine, chin-ups, cable rows). And you should lift stuff (e.g., deadlifts). Again, you need to work hard. You must lift until you can't lift anymore with good form. Start with three sets of 10–12 repetitions. Then, after three weeks or so, increase the weight and move to six or eight sets. Then, after a few weeks, change it up again.

Day 3: Rest or easy aerobic work. If you really love those long, slow runs or bike rides, this is a good day to do one.

Day 4: Medium intervals with circuit work. I use this day to really mix things up. For example, do three longer intervals—say, 30 seconds hard and 30 seconds off—followed immediately by jumping rope for one minute and then a minute of core work. Do the whole routine, the intervals and the mini-circuit, three to four times. And remember, you can do the intervals any way you want: biking, running hills, treadmill, and so forth.

Day 5: Rest.

Day 6: Resistance training. Same as day 2.

Day 7: Long intervals. On this day, stretch the length of your efforts. Your effort should be a bit less intense as compared with days 1 and 4, but keep it up for longer. For example, do a two- to three-minute interval with one or two minutes off. Shoot for six to ten repetitions. Again, this kind of work can be done in a variety of ways. Whatever you enjoy. My wife and I, for example, have a fun Sunday routine that involves biking all the hills near our home.

Days 1–7: Be active! This means commuting to work on foot or a bike, avoiding elevators and escalators, and standing whenever possible (recent research suggests that sitting for long periods of time can be detrimental to your health). When you park your car, look for the spot farthest from the entrance. Instead of sitting in a boardroom, try walking meetings.

You may feel that this is a lot to take on. But if you want to maximize the health benefits associated with fitness, you should strive for this kind of commitment. If you are truly pressed for time, do your best to get in both an interval and resistance workout each and every week. That said, fitness should be a priority, no matter how busy you are.

Even if you find the time to do this kind of intense routine, do *not* expect a quick result. I guarantee that you will feel better (though you are going to be sore!) and that you *will* be healthier. But actual, visible changes take time—perhaps six months to a year. Do not get suckered by quick-fix marketing claims that promise near-instant transformations. That is a recipe for *perceived* failure. Indeed, a colleague of mine at the University of Alberta, Wendy Rodgers, published a study in 2010 that found that even after six months of exercise people's motivation to keep at it does not improve. Rodgers, who is a professor in the faculty of physical education and a renowned expert on the things that motivate us to exercise, studied a cohort of exercisers—some lifelong enthusiasts and some who were trying to get more active—for six months. "The sad truth is that it does not get easier," she told me. "The [fitness] industry promotes appearance outcomes that are impossible to obtain no matter how much you work out. People who are motivated by these goals are doomed to disappointment. In order to stay motivated people must value the activity itself, not the industry [-defined] goals."

Just because you're not morphing into a supermodel doesn't mean you're not gaining real, tangible benefits. Think of fitness as a lifestyle, not as a tool to achieve an aesthetic goal. This is not really what you'd call original advice, but given our society's emphasis on images of physical perfection, it bears repeating.

Nick Wareham and I sat next to each other in the 500-year-old dining hall of Queen's College in the heart of Cambridge University. We were part of a biomedical conference and the ancient room was filled with academics; I decided to stay anyway. The hall has beautiful stained-glass windows and a slight *Harry Potter* vibe. Between courses, speeches, and debates about which sport is more difficult to follow, baseball or cricket (cricket, obviously), I told Wareham about my quest to find the truth about the things that make us healthy.

Wareham is a well-respected scholar (one of my other interviewees called him "a legend"). He is currently the director of the Institute of Metabolic Science at Cambridge and is one of the world's leading experts on a wide range of health and fitness issues. As such, I felt fortunate to have the opportunity to chat, and I may have even fumbled slightly in trying to find the best place to start our discussion. But no prompting was required. Wareham immediately sensed my thesis—that we are bombarded by misleading and untrue health information—and he quickly confirmed what was rapidly becoming the most clear-cut of all the myths: that exercise helps with weight loss. He offered quite a few thoughts as to what forces were twisting the truth.

"In fact, the evidence is better that obesity causes a decline in activity and not the other way around," said Wareham. "The people invested in the science and promotion of exercise don't want to hear this. But the data supports this view. I don't think we should promote things in public health that don't work. Exercise does not work [for weight loss]. There are many, many, many reasons to exercise, but weight loss isn't one of them."

Shortly after I spoke with Wareham, another group in the UK published a study that supports his dinnertime musings. A research team from Plymouth followed 200 kids for 11 years. The results of this study, which were published in the journal *Archives of Disease in Childhood*, suggested that Wareham was precisely right: obesity causes inactivity; inactivity does not cause obesity. Talk about a vicious cycle. Obese kids are not active, have poor body image, and too often adopt a lifestyle that only makes them fatter.

So why are we led to believe so many untruths about exercise, including the idea that it can be used to lose and control weight? Wareham thinks that university-based researchers are part of the problem. Academics in the exercise field want their work to be seen as socially valuable. "People involved in exercise sciences," Wareham told me, "may be inclined to suppress this stuff—perhaps inadvertently—because they want to promote the value of their field and physical activity."

And, naturally, there is a corporate connection. The food industry, especially companies selling products that are calorically and nutritionally problematic, has a strong interest in portraying physical inactivity as the primary cause of obesity. The erroneous belief that we can eat and drink what we want as long as we are active is a win-win for high-calorie food purveyors. It's good news, too, for those branches of the industry that sell fitness products to keep you active. ("Go ahead, have a soft drink after you finish using your Ab-Flexing-Flattener-Machine-Thingy. You've earned it!") Several researchers have studied the strategies of the major food and beverage corporations—sometimes called "Big Food"—and have found that they use sponsorship and public relations tactics similar to those used by the tobacco industry. A 2009 paper by Kelly Brownell and Kenneth Warner, for instance, maps the similarities between the two industries and concludes that one of the current tactics of Big Food is to "emphasize physical activity over diet." The insidious cleverness of it is almost admirable: who could possibly criticize them for telling people to get more exercise?

Wareham provided me with a wonderful example of this tactic. The Coca-Cola company has been a big sponsor of physical activity initiatives. If you go to the company's website you can find the following statement: "We are very involved in supporting physical activity and healthy lifestyles in our communities. Physical activity is vital to the health and well-being of consumers. It is essential in helping to maintain energy balance—the balance between 'calories in' and 'calories out'—for overall fitness and health." The website does not suggest taking the radical steps of eating less and avoiding nutritionally vacant soft drinks.

In May 2010, Coca-Cola was the lead corporate sponsor of the Third International Congress on Physical Activity and Public Health, an event that led to the production of *The Toronto Charter for Physical Activity: A Global Call for Action*. (In the interests of fairness, it should be said that this is an excellent document that notes the myriad benefits associated with exercise and calls for an evidence-based approach. It does, however, also link a lack of physical activity to obesity.) The company has sponsored the building of school exercise facilities and has been closely associated with major sporting events such as the Olympics.

While it is hard to knock Coca-Cola for promoting exercise, it is equally difficult to imagine that the company is not at least partly motivated by the financial upside of sustaining the exercise-as-a-weight-loss-strategy myth. The evidence is clearly telling us that the application of exercise as part of a calorie-in/calorie-out weight-loss strategy, referenced on the company website, is unlikely to work, particularly as a broad public health strategy. We aren't going to get Canada slim by encouraging people to do a bit of moderate exercise, especially if they are "balancing" that moderate activity with a post-exercise Coke.

Reacting to this corporate trend, three Canadian public health experts railed against the market-driven exercise myth in a 2010 editorial published in the journal *Public Health Nutrition:*

> [A] focus on energy-out for energy balance is an inadequate and a potentially dangerous approach, because it is liable to encourage people to ignore or underestimate the greater impact of energy-in. The message put forth can no longer be that of the [Healthy Weight Commitment] Foundation [an industry group]—of which Coca-Cola is a member—which proposes that obesity management and prevention can be accomplished simply by offering "healthy" or "nutritious" alternatives, so long as they are accompanied by enough physical activity for "balance," while avoiding any explicit guidance on energy-in.

I spoke to Sara Kirk, the lead author of the editorial. She believes that the food industry has a subtle yet pervasive influence on public perceptions about the role of exercise in weight loss. Kirk is a professor at Dalhousie University. Much of her work focuses on promoting healthy living, especially around diet and exercise. She echoed much of what I'd heard from others on the topic. "Exercise is not a good way to lose weight," she told me. "The food industry does not want that message to get out. They are about profit and promoting consumption. For example, there's a belief that if your kids run around a bit they can eat a treat. The industry promotes this idea. But even a little treat will blow any [weight control] advantage gained by the exercise."

Another market-driven reason for the twisted weight-loss and -control message: the need to sell exercise gear, classes, and clothes. Everyone wants a miracle. Everyone wants to find the fast and easy path to fitness. The fitness industry leverages our evolutionarily determined predispositions to create the desire for an attractive body and then markets the products and services that will (supposedly) help us reach our tantalizingly always-just-out-of-reach goals. As noted by sociologist Jennifer Smith Maguire in her book *Fit for Consumption: Sociology and the Business of Fitness,* "Fitness media filter the practices and potential benefits of working-out through their vested interests in market expansion, profitability and their legitimation as therapeutic experts."

You should assess every fitness claim—whether it is a new form of exercise or an amazing new product—after first asking yourself this question: does anyone benefit financially from creating the perception that this works? If the answer is yes, adopt a healthy dose of skepticism.

The financial-benefit theme was a big factor for Todd Miller, too, when we spoke in Las Vegas. "The industry rarely sells health. They sell sex. And they sell quick results. And people are driven by results. But, in general, people don't get results—at least that they can see."

Tanya Berry, a colleague of mine at the University of Alberta, has spent her career studying the fitness industry and how people interpret the marketing messages that emanate from it. She agrees with Miller: "The fitness industry is focused on looking good, not getting healthy. In reality, you don't get much weight loss and your looks won't change substantially, at least not in the short term. People try it for a while, but they don't end up looking like [the model on] the cover of a magazine and they quit." Indeed, this cycle of failure is what actually drives the fitness industry. The cultural pressure to look good is intense (it is likely burned into our DNA), and so the industry simply has to find new ways of twisting our instinctive impulses into their evolving business plan. We buy a product. We fail. We move on to another product. And fail. We pay for a one-year membership at a gym. It doesn't work. We move to another activity. We fail again, leaving a trail of money and disappointment in the path behind us.

The businesses that profit from this cycle do not want you to know that the formula for success (meaning true fitness, as opposed to either fitness failure or the appearance of fitness) is really quite straightforward: work hard at fairly simple activities that require few gimmicks or gear.

It's worth pointing out that, as a general rule, all fitness gimmicks are useless or unnecessary. Remember, it is biologically impossible to tighten your tummy or tone particular parts of your body without first losing the fat that surrounds them. Seriously. It is impossible. How your body stores and loses fat is not related to how you exercise particular parts of your body. Yes, you can get stronger in a specific area of the body, but you won't look more toned. One of the ironies of trying to "tone" targeted muscle groups in isolation, without losing the fat around them, is that that region will actually look slightly bigger and bulkier, not tight, toned, or flat. Still, there must be literally thousands of products and exercises based on the fallacious idea that particular bits can be neatly shaped and sharpened.

One of the best examples of the impact of the market on perceptions of the value of fitness activities is the yoga industry. For those of you who are devout yoga practitioners, it might be best to skip ahead a few pages. I am certain that yoga is a fine and fun activity. Many people I know and admire practise yoga.

There is no doubt that yoga has numerous health benefits. For example, studies have shown it can reduce stress, help to alleviate the symptoms of depression, and build strength. But, compared with other activities, its fitness effects are modest. Many yoga devotees view it as more of a spiritual experience or a form of meditation than as a workout; this seems sensible to me. But there are also many more people (in Canada, as of 2005, approximately 1.4 million people practised yoga on a regular basis) who view it as their primary mode of exercise, and looked at this way, yoga is far less effective than many of the alternatives. Is it better than watching TV or sitting in traffic listening to talk radio? Yes. But it doesn't burn many calories (at least compared with more vigorous activities) and it has almost no cardio effect.

It is, at best, a low-intensity resistance workout. As noted in an American Council on Exercise article, "Yoga will improve your strength, but you'll get much stronger, more quickly, by simply lifting weights. As for a cardiovascular workout, yoga isn't the answer for that either." Yes, it results in better flexibility. But, as noted, stretching for stretching's sake has very few health benefits. The flexibility needed for the activities of daily life can be achieved through other, less stretching-focused and more health-promoting exercises. As one of the experts I interviewed put it, "You know what you get good at by doing lots of yoga? Yoga. That's it. Nothing more."

Again, my apologies to all yogis, but think about it: from a physiological perspective, how could yoga possibly be more effective than more vigorous forms of exercise? Your muscles don't know if you're doing the downward-facing dog or a bench press. Whatever challenges your muscles most effectively—thus promoting adaptation—is the better way to go. This is a theme that came up again and again during my discussions with fitness experts. Everyone held the same view, which was nicely summarized by Mark Peterson, whom I spoke to in Las Vegas: "All exercise is about adaptation. Breaking down your muscles or adapting to an aerobic situation. Pick a stimulus that works best. Why use one like yoga, when there's one that is more efficient?"

Why, then, is yoga perceived as such a wonderful fitness activity? One of the most significant reasons is the effectiveness of the marketing that surrounds it. While yoga may feel like an earthy, spiritual, non-materialistic endeavour, it is, in fact, a huge and highly profitable industry. According to a survey done by *Yoga Journal,* Americans are spending close to $6 billion a year on yoga classes and products.

The marketing of yoga is nuanced and rather slippery. It involves the promotion of a lifestyle (which facilitates the selling of clothes) as much as the activity itself. Doing yoga is a form of self-expression, and the market both promotes and capitalizes on this reality. The enormously successful Canadian clothing company Lululemon has demonstrated how this kind of advertising can be disseminated in a manner that doesn't come off as overt and crass. The company, which has sales in the range of hundreds of millions of dollars, uses techniques that some commentators have compared to those used by cults. (A former employee told the *New York Times* it was like working for a "happy cult.")

In Lululemon-speak, customers are referred to as "guests" and salespeople as "educators." Before a new store opens, the company undertakes intensive reconnaissance and sends "missionaries" to local yoga classes to identify instructors who seem likely to be influential. These chosen few become "ambassadors" and get free yoga wear. Of course, they are expected to spread the Lulu word in order to build the base of "Luluheads," a term used by company executives to refer to their devoted "guests."

If you happen to believe that Lululemon is a benevolent entity promoting world harmony through flexibility, with no interest in the non-Zen accumulation of wealth, here, as quoted in *Business Edge News Magazine,* is what founder Chip Wilson thinks of such egalitarian ideals: "Canadians especially, I find, are like a wall of mediocrity. The whole socialist backlog that we've heard for so many years that it's wrong to be rich, that it's wrong to be powerful, that it's wrong to be great, that it's wrong to be an individual—that's just wrong."

Who says money can't buy inner peace?

I am not trying to paint Lululemon as an evil empire. Nor do I want to suggest that they do not make a quality product (on the contrary, I am in full agreement with the hordes of Lulu bloggers who admire the apparel's flattering effect on some of its wearers). Making money is not inherently evil. And using innovative marketing tactics is how companies succeed. This is how the market works. Lululemon has a responsibility to its shareholders to make as much money as possible and it does this, and does it well, by selling the Lululemon philosophy and a belief in the benefits of yoga. But the pervasiveness and effectiveness of these kinds of strategies, coupled with the marketing of all the other yoga products and services, feed the social perception of yoga's health value. We should not forget that this perception is created to make the company money, not to make us healthy.

Of course, I could highlight many other fitness trends. For example, both Todd Miller and Gary O'Donovan believe the same market-motivated approach has created the current interest in core exercises. It's not that we don't need core stability, which, as summarized in a 2008 review article, aids "load balance within the spine, pelvis, and kinetic chain." But they question the value of all the associated products and the emphasis on just "the core." Indeed, O'Donovan believes that much of "this core stability and balance training is garbage." For Todd Miller, these activities are just "fads" driven by the fitness industry.

The evidence on the value of training the core (which is usually thought of as the group of muscles around the trunk) is a bit confusing. While many studies indicate that it helps in athletic performance, I also found research with conclusions like this one from a 2008 study of US Division I football players: "Increases in core strength are not going to contribute significantly to strength and power and should not be the focus of strength and conditioning." In general, it seems the benefits of core stability for the general public, at least from the perspective of injury prevention, are probably real. But the effectiveness of the products and exercises that supposedly focus on the core (I'm talking to you, pilates) is far less certain.

Contrary to the simplistic message found in the popular press and associated with particular workout routines, this is a complex topic, from both a physiological and a fitness perspective. Stuart McGill, director of the Spine Biomechanics Laboratory at the University of Waterloo, is one of the few researchers to study, using scientific techniques to measure muscle use, the best ways to safely train the core. He believes that the popular rhetoric on the subject is confused and uninformed. For example, he notes that "there is no common definition of 'core stability' and, as a result, it is unclear how most core products can actually work." In addition, he says some activities may do more harm than good: "Some products and core exercises, such as some exercises in pilates, incorporate injury-inducing movements in an effort to increase core stability."

And let's not forget the spot reduction fallacy. Core training will not help you lose belly fat. I am certain many people work their core because they believe it will slim their waistline and bring on sexy abs, not because they are driven to improve their "kinetic chain." In any event, there is no doubt that the marketing of stability and core equipment (such as wobble boards, Swiss balls, and Bosu balls) and exercise approaches (such as pilates) has had an impact on the perception of the importance of this kind of exercise that is not in sync with what the actual evidence tells us. Once again, the message has been twisted.

I arrived at Gina Lombardi's gym before she did. Actually, it wasn't *her* gym, but rather the place where she works out. It's an edgy and hardcore martial arts facility just outside Hollywood. Other than a few tough-looking guys—who were kicking, punching, and grunting in a boxing ring—the place was almost empty. When Gina arrived she bowed to the owner of the gym, smiled, and softly uttered some kind of martial arts greeting. She was wearing tight black workout gear and was much tinier than I expected, truly petite, but very clearly highly fit. I suspected I was in the presence of abdominal muscles that satisfied the socially celebrated standard, but I was far too intimidated to ask for confirmation.

Gina Lombardi is part of the culture machine that has set that standard. She is a celebrity trainer. She spends her days crafting the sexy abs we all see on magazine covers, in movies, and on TV. And she is very good at her job; in 2003 she was named fitness trainer of the year. She does frequent TV spots, writes for a variety of fitness publications, and has worked with some of the most famous actors and actresses around. She also, somewhat surprisingly, has a solid reputation in the academic community (which is not generally the case for personal trainers). Indeed, she was recommended to me by several of the researchers I interviewed.

Why, in my hunt for the truth about health and fitness, did I come to Hollywood to see Lombardi? Well, we are constantly told we should strive for a certain kind of physique. I wanted to investigate what the individuals who represent and most famously sell that physique—that is, Hollywood celebrities—do in order to keep fit. I wanted to get a sense of how hard celebrities work to get those sexy abs. After a good deal of email nagging and pleading, I convinced Lombardi to train me like a celebrity. I asked her to pretend that I was a young actor (a significant stretch, sadly) who needed to get fit for a big physical role.

Before I flew to Hollywood, I had a long telephone chat with Lombardi about her work. One of the first things I asked related to a sentiment I have often heard expressed, usually in a bitter tone, by us non-celebrity types: *Well, if I had a personal trainer and hours and hours to train, I'd look better too*. So, was this true? Do celebs really work hard?

"They do an ungodly amount of training," Lombardi exclaimed, with a bit of a chuckle. "Actors have tried everything! This is a huge part of their lives." She told me of one actress (who must remain nameless but who is globally famous and interplanetarily beautiful) who does "two hours of intense, intense work—hard intervals plus weights—every day, seven days a week, all year. Plus, she doesn't eat." Of course, we must keep in mind that this particular actress was born with the biological tools necessary to have a career-creating and envy-inducing body. And yet even with this genetic windfall, she still exercises two hours a day and subsists on a diet one bread stick north of starvation.

During our conversation Lombardi touched on nearly every theme discussed in this chapter. She doesn't think much of yoga or stretching. "I am not a yoga girl. It has its place, but [it's] not a great way to get in shape. In fact, you can't get in shape doing just yoga ... If she [insert name of any beautiful actress here] says she is just doing yoga or pilates she is probably lying."

Lombardi is big on vigorous exercise. "If I had to give one bit of advice: work hard."

She believes in the value of resistance training. "My desert island exercise? Intense resistance training."

She knows you can't lose weight by exercising. "Diet is 80–90 percent of the weight-loss equation."

And she has a list of myths that drive her nuts. "Everyone seems to think that, some day, there is going to be some magic thing—a food, a pill, a drink, or an easy exercise—that will take away the need to work hard and exercise. It is not going to happen."

Lombardi is also keenly aware that she works in a universe that contributes daily to the problem. Celebrity culture helps to build unrealistic expectations that create unattainable goals that are, in turn, used by the fitness industry to turn a profit. In reality, celebrities are the worst fitness role models. For them, achieving a particular look, as opposed to being healthy, is the *sole* goal. "People need to understand that 99 percent of my clients are actors. Most are successful actors. They are highly, highly motivated to stay fit and look good. And most actors are tremendously vain. The profession, the celebrity culture, makes them that way. They want to be thin to look good on camera, and that is the expectation. So I explain what they are going to need to do and what is physically possible. And I tell them that when they aren't on a movie set, they must be working hard."

When I got to the gym and met up with Lombardi, the talking pretty much stopped. I got into my exercise gear and prepared to experience what she defines as "working hard." I will say now that I was skeptical about her ability to push me hard. I have been exercising my whole life. I have competed in various sports at a highly competitive level. In an attempt to rise above my mediocre athletic gifts, I have often exercised like a madman. I have trained with world-class athletes and coaches and have, on innumerable occasions, worked intensely enough to get physically sick. I know, *really* know, what it's like to work hard in a gym. And so, I confess that I expected to enjoy Lombardi's workout. I did not anticipate that I would be fazed by its intensity.

I won't make that mistake again.

Training with Lombardi was, in many ways, the logical end point of my journey into the world of fitness. I had spoken with so many experts that the actual content of Lombardi's workout routine was not a surprise. In fact, her workout was based on the exact activities that my research indicated were the best way to achieve maximum results: intense resistance training, done in a circuit, using exercises that put the body through a wide range of motions. As I said, it wasn't a big surprise.

But the unrelenting nature of her routine came as a shock. It started out okay. The opening circuit, which involved lunges with a weight, planks, jumps, and biceps work (we young actors need good-looking guns), felt moderately intense. I was handling it, no problem. But then when I'd completed that circuit, there was no break. The only rest she offered me was three minutes of vigorous kickboxing. She helped me on with my gloves and encouraged me to start punching immediately. I had never boxed before, so I hit the punching bag like a three-year-old having a tantrum. The look on Lombardi's face was that of a person barely suppressing a laugh.

This cycle continued for an hour. Circuit, then boxing. Circuit, then boxing. My heart rate crept up with each set, and even the simple balance-oriented exercises started to be difficult. By the end I was sneaking looks at the clock and quietly cursing this Lombardi woman. She seemed familiar with this reaction.

"Seriously?" I breathlessly asked her moments after we finished. "This is what your celebrities do every day?"

"Yep, and for some it gets much, much tougher."

After the workout we got in Lombardi's shiny new Mercedes and drove to one of her favourite cafés. En route we had a chat about kids, after-school activities, and some of the more gossipy aspects of her profession. But I was worried that my still-sweaty frame would damage the leather seats, so I found it difficult to focus on the conversation.

By the time we pulled up to the café (valet parking, naturally), I was finally starting to recover. I was also starving. Lombardi recommended the turkey loaf sandwich.

After finishing, I chatted with Lombardi for a few more minutes and then hopped into a cab back to my hotel. My stomach was full of food I wouldn't be allowed to eat if I was about to film a $100-million action adventure. The effects of the Lombardi workout were beginning to hit me. I began to feel groggy, fatigued, and sleepy. I flopped on the bed for a mid-afternoon, post-celebrity-workout nap, thinking as I drifted off how lucky it was that she hadn't subjected me to the tough workout.

2

DIET
THE TRUTH IS SIMPLE—AND HARD

For years it was my job to drive my daughter Alison to her ballet class. It was right after work. Before dinner. Invariably, there was a kerfuffle leaving our house and I never had the foresight to make myself a healthy snack. We'd arrive at the ballet studio, Alison would jump from the car, and I would drive directly to a nearby café for a latte and muffin. I knew muffins were a bad nutritional choice, but I was always famished in that tired-and-lacking-willpower kind of way. So when I first started this routine I asked the young lady working at the café to suggest the healthiest muffin with the lowest calories and fat. The tattooed and pierced waitress—an individual I had unthinkingly decided was a nutrition expert—said something along the lines of "These bran muffins are, like, awesome, and, like, no fat and 150 calories."

The advice seemed sound. She was so obviously fully informed. Thus a ritual was born. I had one of those, like, awesome muffins, as well as a latte, twice a week for two years.

One day there was a new girl working at the café. She didn't know my regular order, so I instructed her to "give me one of those healthy muffins and a latte."

"Are you kidding?!" she said with a laugh. "These muffins aren't healthy!"

To prove her point, she dug up the nutritional information provided by the local bakery that made the muffins. My "healthy" muffin had 750 calories. The average adult should consume between 1800 and 2200 calories a day, give or take. If you added in the latte (probably around 250 calories), my pre-dinner snack contained almost half the energy I needed to run my body for an entire day.

I was stunned, and furious—at myself, at the tattooed muffin expert, at the café, at the bakery, and at those, like, awesome muffins. I was mad because I had eaten thousands and thousands of calories' worth of nutritionally vacant and, in truth, only marginally edible muffins. If I was going to consume 750 meaningless calories I'd rather have taken them in another form. Instead of the "healthy muffin" I could have eaten a medium Dairy Queen Oreo Blizzard (700 calories). Or I could have downed three regular-size bags of peanut M&Ms (250 calories each). Or, using the universal standard for empty calories, a dozen Oreos (55 calories each). Sure, I would have looked odd with a stack of Oreos in front of me, but, from a nutritional perspective, it would have been nearly equivalent to the muffin.

I have often related this story to friends and colleagues, and found that just about everyone has had a similar experience. I'm sure we've all eaten something thinking it was reasonably healthy, only to find out that the food item in question was a caloric avalanche taking place in a nutritional nightmare. But I have often heard the opposite, too. People avoid a food or beverage because they believe it's bad when, in fact, it's a healthy choice. These anecdotes are usually related in a frustrated tone and end with a declaration that there are so many mixed and confusing messages about food that it's impossible to know what to eat. "Screw it" is often the conclusion. "I'll just eat what I want."

Twisted and conflicting messages abound in the world of diet and diets. There is an endless list of trendy diets that promise both weight loss and increased health. Here's a sample of some bestselling titles: *Flat Belly Diet, The Eat Clean Diet, The Kind Diet, Flavor Point Diet, Fiber 35 Diet, The 3-Hour Diet, The 4 Day Diet, The Best Life Diet, The Abs Diet* (no surprise there), and the conspiracy-themed *The Weight Loss Cure "They" Don't Want You to Know About* diet. Governments, health agencies, and research institutions throughout the world have also produced diet guidelines that, while relatively consistent at the core, seem to vary in their details.

And the popular press is a constant source of contradictory advice. Not long ago, my local newspaper, the *Edmonton Journal,* ran a story declaring that bacon was healthy. Yes, bacon! That yummy morning meat fried in its own fat. The headline went so far as to pronounce bacon a "new superfood." I was pretty excited about this perspective-shifting classification. The Caulfield family loves bacon. Our Sunday morning bacon fest could now be enjoyed guilt-free. Alas, just six months later, the *Edmonton Journal* ran another bacon story suggesting that bacon is, basically, poison. Now the paper reported that more than any other meat product, processed food like bacon is associated with heart disease and diabetes and, as a result, should be avoided. Apparently, the American Meat Institute objected to these findings. I wonder why.

Six months, two radically different stories, same meat. So, what is the truth about bacon? And what is the truth about a healthy diet? Why are we fed, so to speak, so many different answers?

Answering these questions formed the next stage in my health-science quest. But I have to start with a warning: diet is such a massive topic—and, as a result of the obesity "epidemic" that has spread across the Western world, one that is increasingly investigated and contested—that I could never hope to cover all the relevant issues in a single chapter.[1] My goal, therefore, is to explore the basic principles of a healthy diet: what and how much we should eat. I want to get to these core truths and identify the primary forces that distort them. To do this, I sought the input of nutritional experts, and then I tried, to the best of my limited self-control, to follow whatever advice they gave me. I wanted to eat the perfect type and amount of food. I became my own guinea pig.

My food odyssey became a personally transforming experience, far more than any other aspect of my investigation into the health sciences. This transformation—which I think was for the better, though I'll allow you to be the judge—was a direct result of understanding the forces that twist the simple, often obscured, truth. In the context of diet, these forces come from within and without. They come from our own misperceptions and from a culture and a massive food marketing machine bent on making us eat as much as possible.

I ordered two pizzas at the café just a few blocks from my office. Two big and delicious pizzas. I was sitting with my personal nutrition expert committee. Let's call them my Food Advisory Team (FAT). FAT is made up of three University of Alberta colleagues who are renowned nutrition experts: Linda McCargar, Rhonda Bell, and Kim Raine. I am fortunate to have such easy access to these individuals, given that all three are respected academics with busy research agendas. McCargar is the director of the Alberta Institute for Human Nutrition and has done cutting-edge research on everything from how to encourage healthy eating habits to the role of obesity in cancer. Bell is also a human nutrition expert. Much of her work focuses on the risk factors, such as poor diet and a lack of physical activity, associated with diabetes. Kim Raine, with whom I've collaborated on numerous projects, is with the university's Centre for Health Promotion and is also the director of the Healthy Alberta Communities project. All three have won many academic awards and know the issues associated with diet, both scientific and social, as well as anyone in the world. Most importantly, however, they were, and thankfully still are, unfailingly good-natured. I don't quite know why they were willing to put up with my crazy "diet project," but they were.

The FAT plan? Simple. To follow their diet advice for three months. I was going to do my best not to let a single, non-FAT-sanctioned food pass my lips.

Given my feeble constitution and my love of sweets, this undertaking would have been difficult at any time of the year. But given that we were proceeding over the summer months—ice cream, barbecues, crass Hollywood blockbuster movies (and the concomitant need for M&Ms), cold beer, kids' parties, vacations—it was clear that it was perhaps a bridge too far. As luck would have it, the FAT summer was also to feature an additional nutrition hurdle, the mother of all diet-destroying distractions: a cruise. In the middle of my diet, I had planned a week-long Alaskan cruise with my family. A week of easy access to beer. A week of all-you-can-eat buffets. A week of poolside reclining chairs ... located close to beer and buffets.

Why, you might ask, would I choose to ruin my summer with a downer of a dietary plan? Well, like exercise, a good diet isn't just advantageous to good health, it's essential. Promoting good nutrition is at the heart of virtually every national and international public health strategy. A bad diet, particularly when extrapolated across an entire population, can have serious societal consequences. Eating a balanced diet reduces the risk of many of the most common diseases, such as cancer, stroke, and heart disease. Living off chips, fast food, soft drinks (and, sadly, M&Ms) can push you a wee bit closer to the grave. The World Health Organization has estimated, for instance, that every year "2.7 million deaths are attributable to low fruit and vegetable intake." And, of course, there are innumerable ailments associated with eating too much; this could very well be the single biggest health issue of the modern age. The US Centers for Disease Control and Prevention (CDC) notes that obesity is linked with type 2 diabetes, numerous cancers (endometrial, breast, and colon), high blood pressure (which has been listed as the number one killer on planet Earth—affecting nearly 1 billion people), liver and gallbladder disease, respiratory problems, sleep issues, degeneration of joints, and infertility.

Which makes it a pretty pertinent question to ask: what, when, and how much should I eat?

When I posed this question to the members of FAT early in our lunch meeting, they each commented upon how amazing it was to them that such a basic human function, eating, has become so complicated. Linda McCargar observed that all we humans really need are "basic nutrients and enough calories to survive, which is obviously not a problem in the modern world. A good diet really is about adequate water, nutrients [protein, vitamins, minerals, fat, carbohydrates], and calories. That's it. But how you get those is the complicated part."

I put it to my esteemed colleagues to get to this "complicated part," but all three members of the FAT were pretty evasive. In fact, it wasn't until we were halfway through our shared pizzas that I put all my cards on the table. "Just tell me," I said to them. "I just want to know what to eat, damn it! Mediterranean? Atkins? Paleolithic? What's the perfect diet? Come on, you guys, it's a simple question. You're the experts. Spit it out ... once you're done chewing ... and give me the goods."

They were hesitant for good reason. It seems that there is no magic formula, no simple list of foods that everyone should eat. Rhonda suggested that "the pattern of eating is key: healthy choices and habits." Okay. Everyone agreed on that. Which meant they all suddenly wanted to know precisely what my diet was made up of at present. They wanted my "dietary habits." They said they needed a "whole assessment" of my nutritional situation. Linda explained that this would require me to keep a "diet diary" and to get my body composition calculated. I told them that the former sounded tedious and the latter humiliating.

Linda laughed at this. She knew that my research for this book involved subjecting myself to a variety of biomedical interventions, many of which seemed more invasive than a calculation of how much flab I carried around. "You have no problem getting your genes tested but you don't want to find out your percentage of body fat?"

I spoke with a mouth full of pizza. "Correct."

The pizza didn't seem so tasty anymore. My task, assigned by the FAT, was to record every calorie I took in, so that they could scrutinize it. I stared down at the last two slices of pizza. "How long do I need to keep this diary?"

Four days, I was told.

I looked again at the slices on the table in front of me. "Including today?"

Yes, they said.

I already hated the process. I pushed the plate away.

With my assessment strategy in place, which FAT decided would include a "before" and "after" calculation of my weight and body composition, we came up with a basic diet and nutrition plan. The first phase would involve the removal of all products (note that I am not using the word "food") that humans simply should not eat. All members of the committee agreed that a healthy diet is about a lifestyle, not a spartan list of super foods. We all know you can eat just about anything, so long as you do it in moderation. But my team also agreed that there were some items that are, as Kim put it, "poison foods"—foods to be completely avoided, products that are not real foods and so should not be factored into the diet plan. After a few weeks of the "no poison" phase, I could then start the "healthy eating" phase. For this part of the experiment I was, according to FAT, going to consume only foods that existing evidence and dietary guidelines recommended.

This was the plan.

As we got ready to leave the café that first day I enthusiastically declared that I was going to get my whole family, including my wife, to follow the diet plan. Why not? I was sure they'd support me.

Kim Raine laughed out loud. "Good luck with that."

Saying so long to the team, I noticed a box under Rhonda's arm. The last two slices of pizza. Damn it.

In the context of diet, the forces that twist the truth reside within us all, especially when it comes to perceptions of height, weight, and what we actually eat. Most of us, particularly men, overestimate our height. And almost all of us—men, women, young, old, short, tall—underestimate how, to be blunt, *fat* we are. A 1998 paper that explored the phenomenon of height inflation noted that self-estimates of height err "in the direction of desired height" because perceptions of height are "colored by [our] dreams." So true. A 2007 review of more than 60 studies found that height was consistently overestimated and weight underestimated. While writing this book I asked many people the average height of a North American male. The most common answer was 5'11". The actual average, according to Statistics Canada, is 5'8½" (this average has been fairly constant over many years, so you can't blame a recent influx of short people from abroad for changing the norm). The study also found that (no surprise here) men under-report their weight by about 4.5 pounds and women by 5.5 pounds. The sad truth is that, if you are a man, you are likely shorter, fatter, and, I suspect, balder than you think you are. (I couldn't find any evidence to support the balder thing, but I sense it must be true. A post-jog burned scalp was my own surprise introduction to the world of thinning hair.)

The perception of weight is, in fact, a complex phenomenon that involves issues of age, gender, and culture. So, I must be careful not to oversimplify issues of body image. For example, a 2010 study found that many healthy-weight female university students perceive themselves as overweight—undoubtedly because of the cultural pressures on women to be thin. But, on a population level, the general trend of overestimation of height and underestimation of weight holds.

This tendency to see ourselves as thinner and taller is a telling comment on our collective desire to project a certain aesthetic, but it also has important implications for public health. A 2008 Statistics Canada study found that if you go by self-reported height and weight, approximately 15 percent of Canadians are obese and 47 percent overweight or obese. If you go by the actual, objectively measured height and weight, the numbers are more depressing: 23 percent of us are obese and 57 percent of us are obese or overweight. This study used 2005 data. Sadly, the number of us who are obese continues to rise. Obesity is an unhealthy or excessive accumulation of fat, usually measured by looking at the weight-for-height measure known as the body mass index (BMI). A BMI greater than 25 is considered overweight and a BMI over 30 is obese. (If you're curious about your BMI, there are many calculators available on the internet. Remember, be honest!) A 2011 study, released jointly by Statistics Canada and the US Centers for Disease Control and Prevention, says 24.1 percent of adults in Canada and 34.4 percent of adult Americans are obese. According to the World Health Organization, 1.4 billion earthlings are overweight and 500 million of us are obese. This is a big problem.

I showed up early at the University of Alberta's Human Nutrition Research Unit to get my body fat composition calculated through a high-tech process called dual-energy X-ray absorptiometry (DXA). There are many other ways to measure body fat, including the old-school skinfold test. However, these methods are not particularly precise. The DXA is considered the gold standard.[2]

The DXA laboratory had the feel of a doctor's office: a waiting room, a place to change, an examination room with the DXA machine. Marcus O'Neill, a fit guy who looked young enough to be an undergraduate, is the assistant director of the unit. We chatted about my diet project as he weighed me and measured my waist and height. Even this simple pre-test process resulted in my first surprise of the day. For most of my life I have told people I was 5'10½". When I was in high school and university, I was, or so my story went, a strapping 5'11". But, according to Marcus, I am just a whisper over 5'9½". Now, to be sympathetic to my memory and vanity, after the age of 40 we humans do shrink at a rate of about 0.4 inches a decade. But this still means that I have never been close to 5'11". I might have touched 5'10"—once.

Then it was on to the body fat.

The DXA machine consisted of an elaborate wand hanging about two feet above an examination table. I lay on the table with nothing on but a lovely light-blue paper gown. A technician strapped me down to prevent movement during the scanning process (which made me wonder if some other examinees had attempted to bolt the room due to fat-panic attacks). The technician started the machine and the wand slowly moved up and down the length of my body, using low-energy X-rays to create a detailed analysis of the percentage of muscle and body fat in each region of the body.

A fat map, in other words. Of me.

I was psychologically prepared to be disappointed with a result of 16 percent body fat, which would have put me in the upper ranges of the "fit" category. As I have noted already, I'm pretty obsessive about exercise. I work out intensely five or six days a week. I use my bike to commute everywhere I go, even in the dead of winter. I pack healthy lunches and snacks. When I travel, I do my best to work out and eat well, and, in order to avoid temptation, I almost always skip conference dinners and receptions. Given my record, I thought I'd have a reasonable body fat count.

Eighteen percent.

That couldn't be, but it was. Eighteen percent.

Now, the fact that 18 percent of my 194-pound body consists of blubber is not horrible compared with the typical 46.5-year-old North American male. For someone of my vintage, I am below average (which, I am told, is about 23 percent) and well within the healthy range (which is between 11 and 23 percent). Indeed, both Marcus and the DXA technician seemed impressed. Thin, young, energetic Marcus said, "Not bad for someone your age." Actually, I'm not sure these were his exact words, but those were the words I heard. An age-qualified quasi-compliment. I'm surprised he didn't pat me on the head and offer me a mug of warm milky tea.

But still, given my extensive efforts to eat well and exercise, and given my (obviously mistaken) belief that I'm still, kind of, sort of—wait a minute, dammit, I am!—an "athlete," 18 percent seemed like a huge number. As I said before: shorter, fatter, balder. Not exactly "higher, faster, stronger," is it? As I left the DXA lab, I thought, "But what more could I possibly do?"

The answer, of course, is simple: eat less. Which brings me to my diet diary. It has been said that systematically monitoring food intake is the cornerstone of weight loss and eating behaviour change. Research has consistently shown that keeping track of what you eat will cause you to eat less and eat better. Using a diary is also positively correlated with weight maintenance. The idea is that, by forcing a degree of self-monitoring, you become more aware of your calorie intake and of the types of foods you're consuming. This works because most humans in the developed world eat much more than we realize, or, at least, more than we are willing to admit to others and ourselves. A well-known 1992 study published in the *New England Journal of Medicine* found that obese patients under-reported their calorie intake by over 15 percent. Subsequent studies have shown that this trend exists in the general population, too. A 2003 study of Brazilian women (not, I am willing to guess, the supermodel sub-population) found that almost 50 percent under-reported calorie intake by 21 percent and that there was a tendency to report the intake "in a socially desirable way" by, for example, "reporting less frequently foods considered unhealthful or fattening, like sweets and fried foods." And a study that looked at the food reporting habits of over 160,000 women (this was part of the big US Women's Health Initiative) found that, on average, reported food intake was 23 percent lower than what this population would need to consume to maintain their reported weight. In other words, these women must have been eating much more than they were willing to admit.

The reason my Food Advisory Team asked me to keep a diet diary had nothing to do with limiting calories; they simply wanted a picture of my eating habits. Yet it had an immediate impact on my consumption. It made me aware of all the "nibbling" I do, especially after work and in the evening. It also made me aware of the muffins, cookies, and bread I buy in the cafés where I sit for hours writing. As a result of the diary, I decided to stop all this extra, unplanned eating. And the result of this small lifestyle adjustment, which I'll detail shortly, was fast and surprising.

After I finished my diet diary—which had to be detailed and include such things as the exact size of my cereal bowl, and not just the number of bananas and homemade cookies I ate, but their precise size, too—I shipped it off to my Food Advisory Team. As it turned out, my intake was reasonably healthy. I had a good distribution between fats, carbs, and protein, though Linda McCargar thought that, given the amount of resistance training I do, I should up the protein.

But, and it's the big but(t), I eat too much. Plain and simple.

During the diary-keeping period I averaged about 2600 calories a day. Since there is no doubt that keeping a diary caused me to eat less, I am guessing that my *real* calorie average would be around 2800 a day. Still, given how active I am, that didn't sound like a huge amount. But this was an incorrect assumption. My FAT suggested that, to *maintain my weight,* I should aim for around 2200–2400 calories a day. Not to lose weight. To maintain my weight.

Incidentally, the immediate effects of self-monitoring offer an insight into why, in the short term, trendy diets work. Following the "raw, blue food diet" or the "butter, fat, and peas diet" forces you to consider what you are putting in your mouth. There is nothing special about the particular foods incorporated into the fad diet. It's the forced self-monitoring that causes you to eat less, which is really the only way to lose weight. Research has shown that other forms of monitoring seem to help too, such as frequently weighing yourself and carefully planning your meals. The point being: tracking what you eat, in whatever way you choose, is likely to help you eat less. If you eat less, you'll lose weight.

I survived the first weeks of my diet plan and the first phase of my project, the elimination of "poison foods," without too many problems. I thought it had gone pretty well, for me, anyway. The rest of my family had, shall we say, mixed feelings. More on that in a moment.

First, what did the FAT categorize as "poison foods"? At our first lunch meeting Kim Raine provided a list of foods that are high in calories and so truly devoid of nutritional value that they simply are not worth eating. "Pop, potato chips, Cheezies, most fruit drinks, sport drinks ... anything that is more manufactured than real." A broad category, for sure, but I got the idea. Avoid junk food, fast food, highly processed products, and sugary beverages. This last category is especially important. Over the past decade, consumption of soft drinks (or, as they say in the US, soda) has increased substantially. Some studies have found that pop accounts for as much as 24 percent of the total calories among 2- to 19-year-olds. And the consumption of pop, which has no nutritional value, is associated with a decrease in the intake of fruits and vegetables. Worse, several studies have shown that pop does not satisfy your hunger and that its consumption has a clear link to the development of metabolic syndrome and type 2 diabetes. In short, pop is nasty. The only beverages you should drink, if you're being strict about it, are water, low-fat milk, and, thank goodness, a bit of alcohol (more on that later). Coffee and tea are good, too. Diet pop, a common choice for those trying to cut down on calories, should also be avoided. To be fair, the research on the health implications of diet pop is far from conclusive. Some studies, however, have shown that the consumption of diet pop is correlated with weight gain and health issues (e.g., diabetes). While the nature of the relationship is not clear—perhaps people simply think they can eat more garbage because they've just consumed a "diet" drink—I vote for erring on the side of caution. No pop of any kind.

I didn't find it too difficult to cut most of the "poison foods," given that I've never been a big consumer of chips or fast food, and I haven't really been a pop drinker for decades. No, it wasn't those items that were the problem. I knew what the problem was going to be. It was going to be one of my staple foods. Part of the Caulfield Food Guide.

M&Ms—at the movies. I'd already been to several films by the time I got a few weeks into my diet plan and already I'd cheated. I had a bag of the little buggers before I started the diet project and it somehow followed me into the movie theatre and sat in the seat next to me. But other than the one moment of weakness, I'd been rock solid.

My kids, on the other hand, were having a rough time of it. If nothing else, the diet plan highlighted the constant exposure children have to junk food. When I first told them of the healthy diet plan, I got "I hate you, Dad!" and "No fair!" and "There is no way I am eating like that!" and "This is your diet, not mine!" and "This stupid diet has made you mean!" I thought it was a bit of an over-reaction. Until we started.

The first day of the project involved a request from my youngest son for a packaged noodle soup product that, according to the label, consisted of the following ingredients: salt, sugar, monosodium glutamate, hydrolyzed plant protein, and many other similar substances. I felt this product qualified as a "poison food," but, using the time-honoured tactics employed by all six-year-olds (volume and argument based on the force of a simple, repeated assertion), my boy insisted on having the soup. I caved, but told him in a firm, authoritative voice, "This will be the last time."

Healthy Diet: 0—Kids: 1

The next day we went to a neighbourhood BBQ. My parenting skills were powerless against the pull of salty snacks and processed meat.

Healthy Diet: 0—Kids: 2

That night the whole family attended a friend's wedding. Free pop!

Healthy Diet: 0—Kids: 3

The Sunday passed fairly well, but Monday involved an end-of-school party. Pizza, chips, and more pop.

Healthy Diet: 0—Kids: 4

The community soccer season also ended that same day, which meant, of course, an evening party involving *more* chips, pop, and processed meat. At least the ice cream sandwiches had some form of dairy-like product in them. I think.

Healthy Diet: 0—Kids: 5

The next night we had friends over for dinner. This resulted in the offering of store-bought ice cream and cookies.

Healthy Diet: 0—Kids: 6

In the morning, the kids headed to their grandparents' cabin at "the lake" (an Alberta euphemism for a shallow body of water surrounded by a suburb of houses and packed with weeds, bugs, and motorized watercraft). A fridge full of pop awaited them.

Healthy Diet: 0—Kids: 7

I showed up at the lake the following afternoon and found our kids eating hot dogs and chips for lunch, which they were happy about, although, to be honest, they still seemed pretty excited about the Fruit Loops they'd had for breakfast. Five minutes into my visit I got dragged into an intense debate with my four-year-old nephew about whether he could have an orange pop. He insisted he hadn't had one yet, but his orange face betrayed him. He switched tactics. Apparently it was now okay for him to have two. His mom intervened and held firm. Finally! One small victory. A victory consisting of limiting a 40-pound four-year-old to one can of syrupy, chemically coloured soda water that contains zero nutritional value and 15 percent of his daily caloric needs.

That night, the kids' dinner consisted of burgers and French fries.

Healthy Diet: 0—Kids: 8

As the summer rolled along, this eating pattern remained constant. There were a few good days when my kids ate mostly healthy, nutritious food, but, in general, the early days of my family diet plan were a disaster. In fact, it was somewhat amazing my kids didn't get scurvy. It must have been the orange pop.

Of course, our society is so permeated with junk food that it has become a social norm. You can't escape the pressure to eat crap. And, from a child's perspective, there is an expectation that junk food should always be available. It's a given. As one of my kids declared in response to my lame attempt to explain the evils of potato chips, "But Dad, bad food just tastes good!" Every "special" occasion in our society is accompanied by salt, sugar, fat, and processed meat. After a game of soccer, all the kids head off to get a Slurpee. After hockey, Timbits and a trip to McDonald's. Even music recitals—and we have been to hundreds to watch our children—have post-performance junk-food smorgasbords.

I do not want to come off as puritanical, or as a sort of diet Nazi. I have always believed that one of the best things about being a kid is getting to eat kid food (if I had a time machine, I'd go back to 1980 and tell the teenage me to *enjoy* that large Skor Blizzard while listening to Blondie's "Call Me" on the lousy Dairy Queen sound system). And as the above account of my family's summer eating habits makes clear, any finger pointing I do should start with the use of a mirror.

But it's hard to deny that our society will call just about anything a "special occasion" just so that we can scarf down some junk food. And there seems little doubt that these diet patterns are associated with the growing obesity problem. Kids under 12 should consume roughly 1400–1600 calories a day. Let's do an extremely rough calorie breakdown of a typical "fun" summer day for a Canadian kid: cereal with milk (200 calories) + juice (100 calories) + two cans of pop (300 calories) + one hot dog on a white-bread bun (300 calories) + the equivalent of a bag of potato chips (200 calories) + a homemade hamburger with cheese and a bun (500 calories) + fries (300 calories) + ice cream cone (200 calories) = 2100 calories. This total does not include any "snacks" and extra chips, hot dogs, cookies, candies, crackers, and burgers the kid might eat (studies have shown that "snacking" has increased substantially over the past few decades, amounting to around 25 percent of total caloric intake). And, obviously, any analysis of the nutritional content of this list of food is going to be gruesome. If you try to cram some fruits and vegetables on top of the menu—a common parental tactic—the calorie count goes even higher.

This example of a single day's menu may be a bit extreme, but I think most of us have witnessed, likely participated in, and almost certainly facilitated a similar kind of childhood gastronomic experience. It happens more often than I'd like to admit with my kids, despite the fact that they're fortunate to be part of a family that is not poor, has some knowledge of nutrition, and lives in a neighbourhood close to stores that provide healthy food options. Many families do not share these circumstances, and that makes healthy choices much more difficult for them. Research has shown that the current obesity dilemma is, to some degree, a socio-economic phenomenon. While the problem of obesity is not restricted to those in the lower socio-economic strata, numerous studies have found that it is more pronounced in lower-income groups. These groups typically have easy access to fast food while healthy options are harder to come by. For example, a 2006 US study found that "the spatial distribution of fast-food restaurants and supermarkets that provide options for meeting recommended dietary intake differed according to racial distribution and poverty rates." A study conducted by a team from the University of Alberta, and co-authored by Kim Raine, found a similar trend in Edmonton. This team came to the conclusion that the odds of exposure to a fast-food outlet "were greater in areas with more Aboriginals, renters, lone parents, low-income households, and public transportation commuters." This situation is surely driven by the economic reality that cheap, easy-access, high-calorie food sells better in certain markets. Calorie-dense food of low nutritional value (sugar, margarine, oil shortening, and mayonnaise) is cheaper than low-calorie food with high nutritional value (such as fresh tomatoes, broccoli, and peppers). A 2005 symposium report by the American Society for Nutritional Sciences that explored "the economics of food choice" found that there is, roughly, an inverse relationship between calories and price. The symposium paper, authored by nutrition scholars Adam Drewnowski and Nicole Darmon, noted that consumers select food based on cost and taste, determinants of food choice that are especially relevant in low-income households. This being the case, it is no surprise that the junk-food market is built on a foundation of sugar and fat.

To make matters worse, we may be biologically predisposed to crave crap. A number of recent animal studies suggest that continual exposure to fast food—that is, high-calorie food with little nutritional value or satiating qualities—can "rewire" our brains so that we become, basically, addicted to the stuff. A 2010 study, for example, investigated how the brains of rats responded to having a palatable, high-fat diet always available. The researchers created the equivalent of a rodent junk-food paradise with scrumptious rat fast food on offer 24 hours a day. The research, which was published in *Nature Neuroscience,* found that an "overconsumption of palatable food triggers addiction-like neuroadaptive responses in brain reward circuits and drives the development of compulsive eating." In other words, the rats became addicted to fast food in much the same way that humans become addicted to drugs. While this neuroscience work remains preliminary, it hints at the possibility that we should think of junk food—or, as my FAT called it, "poison food"—as being in the same category as alcohol, cigarettes, and drugs.

Valerie Taylor, a psychiatry professor at McMaster University and an expert in the field of behavioural neuroscience, thinks this kind of research is central to our understanding of overeating and the place of junk food in our society. I chatted with Taylor on the phone as she dashed between hospitals, yet despite this impressive multi-tasking, she had no problem weighing in with some definitive views on the issue. "Does fast food create an addictive response?" she asked me rhetorically. "I've seen a patient who was so desperate to lose weight that he had his mouth wired shut. He knew that if he didn't stop overeating he was likely going to die. Still he could not stop. He would put his Big Macs in a blender and drink them with a straw. When you see something like that, the word 'addiction' seems appropriate. I just don't know what else to call it."

I'm sure your reaction to that image was the same as mine: how sad, disgusting, and desperate. But was the case she described really the result of a biological addiction to fast food, or was it just a love of food generally?

"Well, I've never seen someone addicted to broccoli," Taylor told me. "The fast-food industry has spent years making food that melts in our mouth and that we can eat quickly so we can go back for more. This food is doing something to the reward pathways in our brain. Remove these stimuli and we crave more."

Not everyone agrees that "junk food" should be viewed in the same way as an illegal drug. Some researchers said it's too early to form a conclusion. Also, some of the experts I talked to thought people should be able to eat the food they love, even junk food, as long as they did it in moderation and balanced it with healthy choices. And categorizing a food as completely off limits may be counterproductive, making it impossible for those losing weight to stay on track. There is at least some evidence that, for young children, restricting access to a particular food may make that food more desirable and so, when they are given access to it, they eat even more than they otherwise would. This "forbidden fries effect" has been observed in several studies. One, published in 1999, concluded: "Restricting access focuses children's attention on restricted foods, while increasing their desire to obtain and consume those foods. Restricting children's access to palatable foods is not an effective means of promoting moderate intake of palatable foods and may encourage the intake of foods that should be limited in the diet."

I agree that a balanced diet is of central importance (more on this below) and that the impact of the messaging that surrounds food choice is tremendously complex. Moreover, I also believe that all diet and nutrition messaging should reflect what the science actually says. Unless the supporting data exists (and, in this context, it does seem to be emerging), we should not use words like "addiction" simply to scare people into healthy eating; this kind of strategy is another of the distortions I am trying to untangle in this book. Nevertheless, I am inclined to agree that, for us grown-ups, fast and junk food should be regarded, as my FAT suggests, as poison. (For kids, the messaging needs to be more nuanced and should, for example, start with parents leading by example.) This is not to say that you can *never* eat your favourite poison, but avoiding it should be the absolute default position. As Taylor told me over the phone, "There is no reason to eat fast food. It serves no purpose."

Well, that's not quite true ... it does make the fast-food industry buckets of money. Which brings me to another reason to think of junk and fast food as poison. We are all fighting (or we should be fighting) the twisting and persuasive forces of a huge and powerful marketing and sales machine. The fast-food industry is engaged in an unending program of trying to make products as available and crave-able as possible. (Big Food invests heavily in the billion-dollar flavour industry in order to procure chemicals that will give their processed food maximum appeal.) In the United States, the fast-food market (which does not even take into account other forms of "poison food," such as grocery store sales of pop and chips) does approximately $150 billion of business every year. It's the industry's corporate responsibility to increase sales to ever-higher levels. Fast-food companies do not care about your health or weight (unless caring will make them money), no matter how cheery, thin, and energetic the beautiful people in their advertising campaigns appear to be. Through a market-driven evolutionary process, and by leveraging our biological needs and desires, fast food has developed into a high-calorie, nutritionally vacant cultural norm. It has permeated almost every nook of our lives. It has, for example, convinced us that every marginally "special" occasion and sporting activity should be accompanied by their products (hence the sponsorship of pastimes and events such as minor hockey, community soccer, and, of course, the Olympics).

From the industry perspective, these brilliant and insidious tactics have been successful beyond belief. Evidence has consistently shown that junk-food advertising has an impact on food preferences, even on kids as young as 18 months. To cite one example, a 2007 study looked at the impact of fast-food branding on children's food preferences. The authors found that simply placing food in a McDonald's container made kids think it tasted better than identical food in a similar but non-branded container. This worked on kids as young as three. Another US study, also published in 2007, found that fast-food chains selectively and aggressively market to particular ethnic groups, such as African Americans and Hispanics. These efforts have resulted in a more positive attitude toward fast food and in increased consumption by both adults and children within these groups. Is it any wonder the fast-food industry spends billions of dollars every year on advertising in the United States alone? It works.

Our exposure to junk-food marketing happens on a scale that is both enormous and shocking. Colourful, bright, and loud, fast-food advertising is a defining element of the North American cultural landscape. When these products and the advertising that goes with them are taken to countries thought to be outside the reach of Ronald McDonald's long arms—countries such as Iraq, China, and India—they are regarded as markers of the successful injection of a Western consumer culture, or, in a word, of wealth. The humorist Ed Druckman once joked that George W. Bush's secretary of defence, Robert Gates, was excited and optimistic about the effect of opening a McDonald's in Iraq: "That is if we can train enough Iraqis to work the shake and fri-o-later machines. But I will say this. We will immediately be delivering outside of The Green Zone."

The ubiquity of junk- and fast-food imagery is especially problematic for kids. For most North American children it forms a central component of their cultural universe. A 2006 US Institute of Medicine report, *Food Marketing to Kids: Threat or Opportunity,* concluded that over $10 billion a year is spent on advertising all types of food to children in the United States alone.[3] And between 1994 and 2004, "the rate of increase in the introduction of new food and beverage products targeted to children and youth substantially outpaced the rate for those targeting the total market." A 2009 survey, conducted by Yale University's Rudd Center for Food Policy and Obesity, found that people greatly underestimated the extent of their exposure to junk-food messaging and overestimated the degree to which healthy food is advertised. The actual figures are staggering. The study reports that "carbonated beverages, fast food restaurants and breakfast cereals spent 18,182 times as much marketing to youth ($1.2 billion) compared to dairy, fruits and vegetables ($66,000 in total)." Survey participants thought the average kid saw one to three junk-food television advertisements a day. The actual number? Almost 15. That equals approximately 5500 yearly television messages about the yummy qualities of salt, sugar, and fat.

Quite apart from the perils of watching television, most people seem completely unaware of the mushrooming scale of advertising on the internet. If a kid uses the internet with any degree of regularity, he or she is likely seeing a considerable amount of cleverly placed junk-food advertising. A June 2010 study out of the University of California, Davis, examined advertisements for internet sites broadcast during children's TV programming on the Cartoon Network and Nickelodeon. They found that 84 percent of the websites—which were based around candy, chips, juice, fast foods, and the like—used video mini-games (the authors call them "advergames") to encourage children to linger. Not surprisingly, the researchers also found that the websites, especially the games, contained a constant barrage of brand identifiers.

I could go on and on about what existing evidence tells us about the horrors of the junk- and fast-food industry, and the harmful impact of its marketing. Admittedly, baby steps have been taken to reverse the social presence of junk food, including successful campaigns to remove it from many Canadian schools. But the multi-billion-dollar junk-food industry is nimble and won't easily give up the opportunity to profit. New corporate strategies will emerge to fill gaps created by any social policy aimed at curbing obesity and encouraging nutritious eating habits. The marketing of junk food as healthy is an example of this kind of misleading repositioning. Many granola bars, for example, have the same number of calories and similar nutrient content as a candy bar. As one nutritional expert was recently quoted as saying, "They're basically cookies masquerading as health food."

So, every time you get a craving for a fried-potato product or a sugar-infused soda water, you should remind yourself that it's being peddled by an industry that is not just selling you a product devoid of nutritional value but actively striving to get you hooked on its food, an industry with a clear corporate strategy to exploit children and our society's most vulnerable populations.

Resistance is not futile. It is essential.

When I made the decision to go on a strict diet it was purely as a means to explore the world of nutrition. I had no intention of losing pounds. I honestly thought I was maintaining an ideal weight, which, it turns out, was a normal, if egotistical, self-perception. Research has shown that most men also think they are better-looking than average. Most of us think we are at least a six or a seven out of ten. A mathematical impossibility, of course. Half of us are uglier than average. There is no other statistical possibility. (Note to math nerds: I realize that the "average" is different from the "median." There could be a group of really ugly men—NHL hockey players, for example—that creates some room for a greater number of good-looking humans. But let's assume looks are distributed normally.)

My DXA moment of truth—shorter-fatter-balder (and probably uglier, though thankfully they didn't have an instrument to measure that)—freaked me out. As a result, losing weight became a secret goal. I didn't even tell my family. If I was to fail, they wouldn't be able to mock me (which is one of my teenage son's primary reasons for getting out of bed every day). But then I started to wonder about weight loss and what society actually notices. In fact, the heck with society. If I were able to lose weight, I thought to myself, would my own family even notice? It became a fascinating little side-study all on its own.

It was a study worth making because the weight-loss program was beginning to take effect. Keeping a diet diary (self-monitoring) and eliminating junk food and mindless snacking, especially pre- and post-dinner, resulted in a loss of about four pounds in two weeks.[4] I was stunned by this, to be truthful, but also secretly thrilled. However, I was alone with the exciting news. No one, not even my wife, Joanne, noticed any changes in my physique. Okay, it was only four pounds, but still—not a peep.

But I forged on, and once I realized I'd lost four pounds just through the habits noted above, I had one paramount thought: could I lose even more?

The next phase of my diet project—the "eat only health food" phase—began with a visit to the office of one of the members of my Food Advisory Team. I showed up late one afternoon and found Rhonda Bell sitting at her desk in one of the University of Alberta's newer facilities, an interdisciplinary centre devoted to diabetes research. I commented upon the nice digs.

"We're pretty lucky over here," Bell said with a smile. "Not quite your typical academic set-up. Don't tell anyone!" We strolled down the hall to her laboratory: it was furnished with benches, bottles, beakers, and high-tech machinery. Just what a lab should look like. There were several graduate students silently tending to their mysterious experiments. Bell took me through some of her current work, which included basic animal studies to investigate the impact of different diets on the pancreas and, in the long term, diabetes. I have known Bell for years and always thought of her more as a nutrition policy expert than a roll-up-your-sleeves bench scientist. I teased her about doing "real experiments," and she laughed at my obvious ignorance about the nature of nutrition research.

The main purpose of my visit was to pick up a food guide put together by several of her graduate students. The guide, designed to help people with diabetes eat a healthy diet, has a clever premise. Bell's students reviewed relevant food guidelines (such as Canada's Food Guide) as well as emerging research on nutrition and came up with a day-to-day diet plan that contains recipes, portion sizes, and general advice about how to eat. All great. But what makes this guide unique and particularly valuable is that all of its advice has a local spin. All the food suggestions can easily be obtained in Edmonton. They even provide shopping suggestions. It's a sensible, "real world" approach.

My strategy for this phase of the diet was straightforward. Eat only the food contained in the guide and in a manner indicated by the best available evidence (e.g., timing of consumption). The FAT agreed that it was a reasonable approach.

I left Bell's office with my food guide in hand, prepared to embark on a journey of deprivation. As we said goodbye, she casually mentioned that she was heading off on a bike trip to France to watch the Tour de France, eat French food, and drink generous portions of fine wine. I decided then and there that she had been delegated by her FAT teammates to be my official tormentor. She was doing a damn good job.

One of the biggest issues associated with the North American diet is portion size. We eat too much. I eat too much. And much of this too-muchness can be attributed to the massive amounts of food we pile on our plates and purchase at restaurants. Portion size is the marketing ploy that plays to our instincts, our weaknesses, our wallets, and even our upbringing.

Let's start with restaurants. It has been noted that in the United States people spend about half of their food budget in restaurants. They consume 20 percent of their meals and 34 percent of their calories outside the home, and most often, about 75 percent of the time, it is a product purchased from a fast-food establishment. In such settings, we consistently overindulge. Why? The answer is easy: we humans are terrible at judging portion size and calories. As noted in a 2008 paper published in the *Journal of the American Medical Association (JAMA),* studies have consistently found that people underestimate the calories in foods, especially high-calorie foods. Usually their guesses are off by nearly 100 percent. Yes, 100 percent. In other words, if you think something is 400 calories, it is more likely to be 800. For some foods, such as fettuccine alfredo or chicken fajitas, people underestimate the "caloric content by 463 to 956 calories."

As a general rule, you should view all restaurant food as high-calorie. This is especially true of fast food. Remember, most adults need between 1800 and 2200 calories a day. According to nutritional information provided by McDonald's, a Double Quarter Pounder with Cheese is 740 calories. A side of large fries is 500 calories. Add a 1100-calorie large Triple Thick chocolate shake and an apple pie thingy (250 calories) and you have more calories than you need for the entire day—minus most of the needed nutrients. If you were a kid, this meal would be much, much more than you require.

We also typically underestimate our intake at more upscale restaurants. During the four days that I was keeping my diet diary I travelled to Vancouver for work. As always, I tried to eat in a fairly healthy manner. So, for dinner, I went to a trendy, mid-level restaurant with good-quality food offerings. I ordered a salad with roast chicken and what I thought was a vinaigrette dressing. The young waitress gave me a hard time about my "healthy" choice.

She said something like, "That salad will leave you room for an extra beer!" Giggle. On cue, I also ordered a beer. Turns out the salad was over 900 calories. The pint of beer added another 200 plus. Throw in my breakfast, lunch, and some bad plane food, and I consumed almost 4000 calories that day.

Four days of a diet diary, just four days, and I was starting to get a clear picture of how body fat accumulates.

My lack of caloric self-awareness is hardly unique. Indeed, ignorance about how many calories we need and how many we consume seems to be the norm. A 2010 survey of more than 1000 Americans conducted by the International Food Information Council Foundation found that only 12 percent of respondents "can accurately estimate the number of calories they should consume in a day for a person their age, height, weight, and physical activity." In addition, "of those who say they are trying to lose or maintain weight, only 19 percent say they are keeping track of calories."

We are bad at estimating both the number of calories in food *and* how much food we need. Not a great combination, unless you are in the food industry. To make matters worse, we may also fool ourselves into believing that all this calorie counting, overeating, and obesity isn't relevant to us. The authors of the *JAMA* article noted above, Mark Berman and Risa Lavizzo-Mourey from Princeton's Robert Wood Johnson Foundation, suggest that the tendency of individuals to perceive themselves "at lower risk for adverse events than the average person," a phenomenon called "optimistic bias," is partly to blame for our inability to control caloric intake. In other words, we all think that getting fat and unhealthy is going to happen to the other guy, not us. The optimistic bias tricks us into eating more because we think, incorrectly, that the risks associated with the consumption of excess calories are not a significant personal issue.

Another reason we overindulge is that we all love a good bargain. Do you think restaurants might have figured this out? Up until the 1970s, the fast-food industry ballooned at an ever-accelerating pace. It grew and grew to meet an endlessly expanding appetite for junk. In the 1980s, however, the market started to become saturated and the competition for customers became intense. As a former CEO of McDonald's, Mike Quinlan, wrote in a 1998 *New York Times* article, "You have to take away business today in order to grow." One of the tactics adopted by the industry to take business away from competitors has been to offer bigger portions. Huge portions. Some fast-food items are two to five times larger than they were when originally introduced. It's not hard to do because, as noted earlier, fast food is made up of relatively cheap ingredients. Profit margins can remain high even when the products sold are super-sized. (It is often reported that US farm subsidies, which started around the 1970s, promoted the overproduction of cheap food, thus facilitating the peddling of huge quantities of low-quality fast food. While this thesis is logical, the data supporting it is, to date, somewhat equivocal. A 2007 economic analysis out of UC Davis found little connection between the farm subsidies and obesity. Ditto a 2009 study. Regardless, for myriad reasons, many of the ingredients that are essential to fast food are cheap, thus allowing for the profitable production of massive portions.)

Big portions have become a North American norm. And research has consistently shown that we have embraced the norm. If a plate piled high with deliciously empty calories is placed before us, we almost always cram them down, even if we are not hungry or do not particularly like what we're eating. A clever study by Brian Wansink and Junyong Kim, two food-behaviour experts from Cornell University, demonstrated the degree to which portion size influences our eating habits. They gave more than 150 moviegoers free popcorn. Some of the moviegoers got fresh popcorn and some got stale, 14-day-old popcorn. The researchers found that simply serving the popcorn in a bigger container resulted in more consumption, even when the popcorn tasted bad.

Moviegoers who were given a big container ate 45 percent more of the fresh popcorn and, astonishingly, 34 percent more of the stale stuff. Wansink's team performed another ingenious experiment. They fed a group of research participants from normal soup bowls and another group from bowls that were subtly and secretly self-refilling (the soup was pumped in from below the bowl). Those who ate from the self-filling bowls consumed a whopping 73 percent more soup, but they did not believe they had consumed more, "nor did they perceive themselves as more sated than those eating from normal bowls."

Innumerable studies have confirmed this calorie-cramming tendency, and many others have speculated about the connection between the growth in portion sizes and in our waistlines. For example, in countries where more reasonable portion sizes are the rule, such as in France, the obesity rates are much lower. (This has been offered as one explanation for the finding that the French, who eat rich food, are skinnier than North Americans—a phenomenon often called the "French paradox.")

In response to growing criticism about portion size, some fast-food companies eliminated their largest offerings. For example, as a result of the attention generated by the documentary film *Super Size Me,* McDonald's phased out its largest fries and pop. The current versions are still huge, of course. In fast-food restaurants "huge" is a consistent, unrelenting theme. In fact, the most recent marketing trend appears to be a return to a full embrace of the disgustingly massive Wendy's Baconator Triple, which was introduced in 2007, features two meat patties, two cheese slices, and bacon and clocks in at 1350 calories. Think about that. Nearly 1400 calories, for a lunch sandwich. Burger King has the Triple Whopper with Cheese (1230 calories) and the Quad BK Stacker (930 calories). A&W's Chubby Chicken Dinner contains 1230 calories and IHOP's 2010 creation, the Pancake Stacker combo, has 1250 calories. And the Ultimate Grilled Cheese Burger Melt, released in the summer of 2010 by the US chain Friendly's, appears to be the current champ of the fast-food calories arms race, at 1500 calories. And remember, to maintain your weight you should be aiming for between 1800 and 2200 *a day*.

The manner of our upbringing is usually another factor in the portion size dilemma. Many of us were raised with a "clean your plate" mentality. We were told that we could not have the delectable dessert until we finished all the food on our plate. Turns out that this is a terrible message. It teaches kids to ignore internal cues regarding hunger and satiation. It encourages overeating. A kid who stuffs himself is rewarded by more yummy food. As adults, we fall back on this training whenever we go to a restaurant: we eat everything on our plate even if we are full halfway through.

A study published in 2006 examined four different approaches to parenting and found that an "authoritarian" style—the classic "clean your plate" approach—was the one most closely related with overeating and obesity. A more flexible, or even "permissive," approach correlated with a healthier attitude to food. The study concluded that strict mothers were nearly five times more likely to raise overweight first-graders than mothers who treated their children with flexibility and respect. This doesn't mean that you should let kids eat what they want, but rather that strict parenting was found to be even worse, from an obesity point of view, than "negligent" parenting. The most effective approach, called *authoritative,* involves clear rules, but incorporates healthy choices and a degree of flexibility. We want to teach kids to enjoy a healthy amount of healthy food.

In any case, as a result of our love of a bargain, our inability to accurately estimate how many calories we need and consume, our optimistic bias, our "clean your plate" mentality, and our exposure to countless environmental cues and intense marketing pressure to eat, eat, *eat,* we tend to consume whatever is put in front of us.

To combat this tendency, I formulated the following restaurant rules as part of my diet plan:

1. Stay out of fast-food restaurants. Obviously.
2. Have a plan in place before you go into a restaurant and stick to it. ("I am ordering a starter salad and a half-pint of beer and that is it.")
3. Be suspicious of all sauces and salad dressings. (If possible, get them on the side.)
4. If you must order an entree, try to eat only half. (Take the rest home in a doggy bag or simply leave it on the plate. What's worse: waste on the plate or waste in your stomach?)
5. Completely ignore what your dinner companions are ordering. (Research has shown that the eating habits of friends and family have a big impact on our behaviour.)
6. If you order something healthy and low-calorie, *do not* balance it with something unhealthy. If you order a nutritious, low-calorie entree do not think you have earned the right to top it off with a piece of cheesecake or a side of fries.
7. Do *not* compromise on Rule 6.

Some observers think that the widespread delusion that it's okay to balance a low-calorie dish with something obviously fattening explains why the proliferation of healthy options at restaurants has had no impact on obesity rates. Many restaurants, including McDonald's, now have reasonably healthy, low-calorie offerings. But, predictably, these new menu items serve more as a strategy to get people in the door than as a way to actually alter consumption habits. Studies have shown that when people buy a healthy item they often underestimate the calories it contains and then, compounding the problem, buy high-calorie side dishes. A 2007 study explored the effect of "health halos," the phrase the authors of the study used to describe the effect, and found that it is a remarkably strong force. The authors reported some astounding, and depressing, results: "We find that consumers chose beverages, side dishes, and desserts containing up to 131 percent more calories when the main course was positioned as 'healthy' compared to when it was not—even though the 'healthy' main course already contained 50 percent more calories than the [comparable] 'unhealthy' one." Another 2009 study, out of the University of Toronto, found that people eat about 35 percent more when they think a food is healthy, which is problematic on two levels: first, even very healthy foods (such as almonds) can be high in calories; and second, food marketed as "healthy" frequently isn't. A 2010 study found a similar effect when the label is "organic." People tend to view organic as healthy and therefore believe they can eat more of it. And this leads me to my last restaurant rule:

8. Do your best to investigate the actual calorie and nutritional content of your favourite restaurant meals.[5]

There are many good calorie-counting websites and phone apps. I found that after a couple of weeks of doing this kind of background research I got much better at estimating the potential damage. Here let me offer you a warning to protect your social life: *keep the calorie counting to yourself.* It is my experience that telling people how many calories they're eating, especially while they're enjoying a meal, is not appreciated. I have gotten a few "don't ruin this for me" looks, and, from my lovely wife, the romantic dinnertime admonition: "I don't want to play your little diet game." In fact, it's best to keep your entire diet and/or weight-loss plan as private as possible. One night my wife and I went to dinner with friends and the entire table became enraged when I didn't eat my share of a pizza entree. It was as if I'd broken a sacred social contract. It didn't matter that I just wasn't hungry. "Call us when your diet project is done," said one of our friends. Everyone nodded in agreement. I was the group downer.

These rules may seem intense. You may feel they should be linked only to an austere weight-loss program. In fact, they should be part of everyone's long-term lifestyle strategy to *maintain* weight and battle the social forces pushing us all toward pudginess. If you eat out frequently, which I do, and if you travel regularly, which I do, these rules should be viewed as a new, enduring routine. Because, let's face it, we're creatures of habit, whether we're conscious of it or not. All I'm talking about is substituting one routine, a healthy one, for another, unhealthy one.

As my weight-loss experiment continued, the results spoke for themselves. I kept shedding pounds that I'd previously deluded myself into thinking I did not need to lose. After about a month on my healthy diet plan, I'd lost nearly ten pounds. This was a revelation to me, but make no mistake, it was a tough slog. I was hungry ... *all the time.* Often grumpy.[6] I seriously craved cookies and my beloved M&Ms. At the start of the program I would wake up in the middle of the night with hunger pangs, often after a dream about food. In one dream I purchased M&Ms at a crazy all-night convenience store with weird multi-coloured lighting, a freakish sales clerk, and an uneven floor. When I opened the bag, the M&Ms were dust. Call up a Freudian therapist if you want to figure that out, but it sure felt cruel inside the dream. More than once during this phase, I found myself trying to imagine how incredibly hard it would be for someone who *had* to drop a large amount of weight for health reasons.

Then an amazing thing happened at the ten-pound mark. Joanne noticed. I suppose I might have planted a few seeds, but although she didn't comment directly on my physical appearance, she asked me, "How's the healthy diet going? Have you lost weight?" Up to that point no one else had noticed, or at least no one had said anything. Not my brothers, my kids, my work colleagues, my students, or the guy who sells me my coffee every morning. The reason for the silence might very well have been that I am not worthy of a compliment, but for the purposes of this book and until proven otherwise, I have adopted the no-one-noticed explanation.

I had lost ten pounds, and although I had dropped most of

the "poison foods" from my diet, there was really just one reason for my weight loss: portion size. For decades I had been feeding myself too much food, particularly at dinnertime. As noted many times in this book, we humans do not need many calories. We have evolved to be highly efficient eating machines, a reality that some public health officials are starting to emphasize. In 2010 the New York City Department of Health started a poster campaign to encourage New Yorkers to eat less. Across the top, in bright letters, the posters read, "2000 Calories a Day Is All Most Adults Should Eat." The posters also have a picture of some food item with a little flag displaying the calories it contains. A chicken burrito lunch, for example, has a flag telling us it contains 1200 calories. The poster asks, "If this is lunch, is there room for dinner?"

I did my best to follow the portion size recommendations found in Rhonda Bell's food guide, as well as other sources, such as Canada's Food Guide. As noted, my Food Advisory Team suggested that, for me, 2400 calories would be a good weight maintenance goal. But given my new desire to slim down, I decided to aim for around 2000–2200 calories a day, using portion size as the primary reduction mechanism. Smaller lunches and smaller dinners.

Making a quick assessment of portion size is not easy. For example, there is a great deal of confusion about the difference between *portion* size, which is the quantity of food on your plate, and *serving* size, which is the unit most often used in food guides (and which is, therefore, somewhat constant). A *portion* may contain several *servings*. In addition, store-bought products, like cereals, chips, and pop, often use the serving size as the unit for the nutritional information on the package. But, more often than not, the package contains far more than one serving. The food industry uses this as a way to create another subtle twist in the information you get. At first glance, the nutritional label on the side of a 42-gram (1.5 ounce) bag of chips may seem to say it contains 155 calories. But this calculation may be for a particular serving size, say 28.3 grams (1 ounce), not the entire bag. How often do you eat two-thirds of a bag? To make matters more confusing, the serving sizes found on nutrition labels do not necessarily correspond with the serving sizes recommended by the Canadian or American food guides.

Portions, servings, calories. Grams, ounces, fractions. Confusion abounds.

"Portion size is fundamental to healthy eating," says Geoff Ball, a nutrition expert and University of Alberta professor. "And understanding portions and serving sizes is tough. Portions vary depending on who is serving, and studies have shown that people are very bad at estimating the size of portions. Everything affects our perceptions—including colour, shapes, and context. It is tougher to guess the servings in an amorphous blob of food than a piece of fruit."

Ball is also the director of Alberta Health's Pediatric Centre for Weight and Health. The centre's mandate is to provide weight management advice to overweight and obese children, youth, and their families. And, it turns out, a significant aspect of that mandate is teaching appropriate portion size. "Portion size is one of the biggest issues we deal with," Ball told me as he showed me around the centre, which is located in an old hospital in downtown Edmonton. "Many of the families I work with believe that their kids' weight problems are medical or biological. Actually, biological reasons for obesity are extremely rare. I've probably assessed over 400 kids and no kid has had a thyroid problem, which people commonly believe is the problem. It's a lifestyle problem. Parents also think their kid will grow out of it. I don't know why they believe this. Wishful thinking, maybe. But this isn't rocket science. They need to see it as a long-term lifestyle issue associated with the kind and amount of food they eat."

The centre provides advice on how to assess portion size. They often compare familiar objects like hockey pucks, baseballs, and tennis balls to quantities of food (a hockey puck, for example, is roughly ½ cup and a baseball about 1 cup). Ball's team also provides classes on healthy eating. He told me that "by far, the portion size class is, for the families, the most surprising." For the class, Ball cooks up a big pot of pasta and asks people in the class to dish up what they would normally eat for dinner. The participants come forward and fill a plate with pasta. "Usually it's a big mound of food. It often amounts to more servings of grains than needed in a whole day."

How much, then, do we need? What is an appropriate serving size?

For the purposes of this discussion, let's use the recommendations from Canada's Food Guide for an average adult male (these are quite similar to those found in the US guidelines). It recommends around nine servings of fruit and vegetables, eight servings of grain products, two servings of milk products, and three servings of meat or a meat alternative (the amounts vary by age and sex). How much is this? One serving of fruit is, roughly, one piece of fruit. One serving of uncooked leafy vegetables is equal to one cup. A serving of grains equals one slice of bread or half a cup of cooked pasta. A serving of a milk product is one cup of milk or ¾ of a cup of yogurt. And one serving of meat is, for example, about a 3-ounce breast of chicken, which is not much. It is about the size of a deck of cards.

The students in Ball's class not infrequently fill a big plate with pasta—the equivalent of five to seven servings, depending on how high they pile it. As Ball noted, that is a lot of food. If you had cereal for breakfast (one to two servings) and one sandwich for lunch (two servings), you haven't left any room for a huge pile of pasta, especially if you are a kid (Canada's Food Guide recommends four to six servings for those under 13).

Of course, you must also consider calories when you're reckoning portion size. It is possible to eat according to the food guide suggestions and still consume more than your recommended daily calories. For example: four bagels could amount to 1200 calories; nine large bananas is around 900; a cup and a half of yogurt might have up to 250 calories; and six large eggs 350. While this would be a weird day's menu, it technically meets the serving size recommendations but clocks in at around 2700 calories, more than most of us need (and this is without anything to drink). Indeed, some obesity experts feel that following the guide could lead to weight gain, especially when you consider the additional "discretionary calories" we all, inevitably, eat (such as the occasional cookie, ice cream, or French fries).

The distinguished Harvard nutrition expert Walter Willett agrees that "this is a serious problem." I corresponded with Willett about the concerns associated with guideline recommendations and overeating. "People have been told to eat more of many foods," he wrote to me in an email, "including some good ones, but there [is] no message to eat less of anything ... we can see the problem."

As with the new American guideline, which was released in early 2011, the 2007 Canada's Food Guide does recommend, in small print near the end of the document, "limiting foods and beverages high in calories, fat, sugar and salt." But Willett finds this kind of advice to be "pretty obscure." In fact, here in Canada, Yoni Freedhoff, the director of Ottawa's Bariatric Medical Institute, has been quoted as suggesting that the guide is "obesogenic" because it promotes the consumption of too much food and provides insufficient guidance on calorie intake. Valerie Taylor, not one to hold back a controversial opinion, feels even more strongly: "The intakes are crazy," she told me. "No one can eat that much." I asked her if she thought the food industry had influenced the serving recommendations. "Absolutely. Big Food is like Big Pharma or Big Tobacco."

This statement deserves comment. The issue of corporate and industrial influence on what we are told about our health is, of course, a theme of this book. And many have speculated that the food industry has had a great, twisting influence on food policy and the production of nutrition guidelines, particularly in the United States. While the remedies chapter will discuss industry's influence on knowledge production, it's hard to deny that it has also played a less than constructive role in food policy and nutrition research. We have seen how much the food industry has shaped our eating habits through marketing strategies and the sale of fast and junk foods. In the fitness chapter, I outlined how industry has promoted an inaccurate view of the role of exercise in weight loss—the misleading exercise-so-you-can-eat proposition—to market their products. But has the food industry influenced national food policy and, by extension, what we are told about nutrition by government? Specifically, was the number of recommended servings influenced by industry? Given the amount of money at stake, it would be naive to assume that they haven't been part of the story. For example, it has been estimated that in 2006 beef production contributed $26 billion to Canada's economy. If the government recommends that we eat less beef, a point recommended by some nutrition experts, it would have a direct impact on the beef producers' bottom line.

It is widely accepted that early versions of the US food guide favoured industry interests to the detriment of public health. Numerous scholars, most notably New York University's Marion Nestle, have documented the powerful role of the food industry in this context. For example, in a 1993 article, Nestle reported how lobbying from meat producers resulted in federal dietary advice that evolved "from 'decrease consumption of meat' to 'have two or three (daily) servings.'" Such stories stand as important cautionary tales and should force us to critically appraise the role of industry in the development of current guidelines. Past problems notwithstanding, it seems that there is a bit more confidence that the current version of the US food guide is based on good science. Walter Willett believes that industry influence on the background work for the current version was minimal. It was informed by good science. But he recently warned in the Harvard School of Public Health's online publication *Nutrition Source* that the final version of the guidelines still has an industry-friendly spin. Willett speculates, for example, that "the continued failure [of the guidelines] to highlight the need to cut back on red meat and limit most dairy products suggests that 'Big Beef' and 'Big Dairy' retain their strong influence within this department [the US Department of Agriculture]." This happens because the scientists and nutritionists who worked on early drafts don't have a say on the wording of the final product.

In Canada, the current guidelines were developed with input from the private sector. Members of the Food Guide Advisory Committee, the 12-member entity responsible for the revisions, included policy-makers and independent academics, but also representatives from the Food and Consumer Products Manufacturers of Canada, the Vegetable Oil Industry Council, and the BC Dairy Foundation. It's hard to imagine that these organizations were there to champion the simple goal of lower food consumption.

While I sympathize with the idea of engaging industry in the development of a national nutrition strategy—industry makes and distributes most of our food, after all—I doubt the wisdom of including them in the actual process of crafting guidelines, which should be based entirely on the best available evidence. We cannot forget that industry's goal is, inevitably, profit. Consequently, conflict is unavoidable. Marion Nestle and David Ludwig drew attention to the issue in a 2008 commentary: "In a market-driven economy industry tends to act opportunistically in the interests of maximizing profit. Problems arise when society fails to perceive this situation accurately." Given this conflict, industry's direct involvement in the development of food guidelines will lead to a perception of bias and a decrease in public trust or, at worst, a distortion of the scientific recommendations in a manner that serves industry. Indeed, there are innumerable examples of the ability of industry to bend conclusions to their benefit. Evidence published in 2007 in the journal *Public Library of Science Medicine* found that industry sponsorship of nutrition-related research resulted in conclusions that were biased "in favor of sponsors' products, with potentially significant implications for public health."

What is the situation in Canada? There seems no doubt that industry involvement in the work of the committee has resulted in a degree of mistrust, as highlighted by my discussions with Valerie Taylor. And there are indications that the industry can have a direct impact on the evidence used to shape policy, resulting in a more insidious problem: can we have confidence in the research that informed the guidelines? The answer is a qualified yes. With some exceptions, the experts I spoke with felt that, from a nutrition perspective, the guidelines in both Canada and the United States provide sound nutrition advice, even if it's slightly flawed in the details. Despite the possibly inappropriate role of industry, in other words, existing evidence supports the general themes presented in the guidelines. Yes, there are issues, such as confusion about how much food should be eaten and some quibbles about the recommended number of servings (Walter Willett, for example, believes that the evidence does not justify more than a moderate amount of milk consumption, perhaps one serving of milk a day) and concern that there is not enough emphasis on the "eat less" message (in one of her critiques of the US guidelines, Marion Nestle has noted that "eat less is not nice for the food industry"). But as an outline for the basics of a balanced diet, the guidelines are helpful and in line with the major themes that decades of nutrition research have developed.

All three members of my Food Advisory Team agreed that the guidelines were essentially sound. Linda McCargar, for example, is satisfied with the process that led to Canada's Food Guide (and she has been involved in the development of national nutrition policy). She acknowledges that a casual reading of the recommendations could lead to overeating, but that that's largely a matter of interpretation. "The food guide is first and foremost to assure adequate nutrition, intake of all the nutrients," she told me. "I would agree that servings must be chosen carefully to be able to keep energy intake at a level to maintain weight. However, the servings in the food guide are relatively small. The portion sizes that are confusing are the ones available in our food supply—for example, a 12-ounce bottle of orange juice is three servings. Two cups of rice in a restaurant is equal to four servings. I think it's this translation that gets confusing."

Many other experts expressed a similar view. Yes, we need to be vigilant to ensure that industry does not inappropriately influence policy. But the existing guidelines are solid. Diane Finegood, a professor at Simon Fraser University and former scientific director of the Canadian Institutes of Health Research's Institute of Nutrition, Metabolism, and Diabetes, told me that the process that led to the Canadian guidelines was appropriately transparent and informed. "The best available evidence was used to produce the right balance and mix of food," she said. "One of the problems is that all food guidelines are often taken out of context. People want them to do more work than they can do. Developing a single set of guidelines for a diverse population is a real challenge. People should simply view it as a general guide about how to eat."

And now, the heart of the matter. *How* should we eat?

This question was relevant to my diet. Two months into it, I had dropped 14 pounds. A few more people had noticed, including my mother-in-law, who called me "skinny," which I decidedly am not. *Skinny* is a word that has not been associated with my physique in more than two and a half decades. True, my pants didn't fit. And my favourite T-shirts were all baggy. But I felt great ... except for the constant hunger and cravings for cookies and chocolate-covered peanuts. The diet-destroying Alaskan cruise was looming like, well, like a cruise ship on the horizon, but so far so good.

Not only that, but for the second month of the first two, my wife had decided to join in the "healthy diet" and had lost five pounds. The diet had also become a household project. Joanne and the kids made numerous meals from Rhonda's food guide, and every child was sent off with "FAT-approved" lunches for their summer camps. It was still difficult to adhere to the diet, make no mistake, but having a gang on board helped. The best thing about it was that the kids were finally starting to eat healthier foods. And, although I may be fantasizing about this, they actually seemed to be enjoying it. To be more precise, they seemed to not even notice the shift. We were publicly focusing only on the avoidance of "poison food," and their sole job was to recognize and limit their consumption of junk food. We, the meal-producing parents, did the rest.

So, what *were* the Caulfields eating? We kept the plan simple. For me, it has not been a "diet" so much as a pattern of eating, which is what my FAT had suggested at our first meeting. I had truly begun to believe and embrace those clichés about healthy eating and dieting, the biggest of all, perhaps, being that it's a lifestyle, not a magic formula.

Here is my non-magic formula.

I will start with the stripped-down basics, a one-sentence summary that will take you 90 percent of the way to a healthy diet. It echoes what many other diet commentators have proposed. It's not perfect, but it's a rough rule that will make a serious difference for most people. And, for most, it will likely result in improved health and weight loss.

Eat small portion sizes, no junk food, and make sure that 50 percent of what goes in your mouth is a real fruit or vegetable.

Here are five steps that will give this one-sentence formula more teeth:

1. *Accept that 1800–2200 calories a day is not a lot of food.* It really isn't. Sorry. It's three small meals and a few healthy snacks. Use a diet diary for a few days to figure out how many calories you are eating. This takes a bit of work, but it is essential. It provides an opportunity both to learn about the calorie content of common foods and to gain an understanding of your pattern of eating. Be brutally honest. Unless you're an alien robot or have a family of tapeworms living in your gut, eating more than the 1800–2200 calorie allocation will, inevitably, result in unhealthy weight gain. Ideally, you should take some time to investigate exactly how many calories you actually need to meet your goals (this requires keeping a weight and diet diary over a longer period of time). Every human is slightly different, with a great emphasis on *slightly*.[7] As we get older, our calorie needs decrease. And most active teenage boys (damn them!) can consume around 2500–2700 a day. But for most adults, the variance in how much we burn is likely, at the very most, a few hundred calories a day. (This relates to the myth about fast and slow metabolisms, to be touched on below.) For this step, portion size is likely your biggest challenge. And do not make the mistake, as I did, of allowing yourself to consume more because you exercise frequently. As numerous experts have told me, you can't outrun a bad diet, figuratively or literally.
2. *Get a sense of what a healthy diet looks like.* To do this, you can turn to the national food guides discussed above. There are also a number of fascinating studies, to be touched on below, that highlight some especially healthy dietary practices. But despite all the debates about nutrition policy, all the media coverage about various super foods, all the ongoing research into the details of nutrition, and all the innumerable fad diets, the basic tenets of healthy eating are ridiculously simple: about 50 percent of what goes in your mouth should be a real fruit or a real vegetable. About 25 percent of your food should be a grain product, and, whenever possible, this should be

something with whole, unprocessed grain. The remaining 25 percent should include some lean meat (or a meat substitute), eggs, and/or low-fat dairy. And always, always avoid all junk and fast food.

3. *Fit the healthy foods into your daily calorie allocation in a way that works for your lifestyle.* This is, in some ways, the most difficult part. This is where the Canada's Food Guide serving sizes are meant to be helpful. But, to be frank, despite months of trying, I couldn't wrap my head around the serving size approach. It just does not fit with how I live day to day, meal to meal. When I look at a meal, I find it difficult to convert the portions into serving sizes. Instead, I think of my daily intake as one big plate. On that plate I can put about 2000 calories of food. Half the plate should have fruits and vegetables. The rest of the plate should be 25 percent grains and 25 percent dairy, meat, and eggs. Every day does not have to have a perfect allocation. But this should be the long-term average. After you have kept a diet diary for a few days, you will get a sense of what your plate should look like and what you need to adjust. You will also find that a vast number of fantastic, flavourful, and filling foods fit on this plate. It is not a completely spartan and depressing menu, just a different one.
4. *Be conscious of all the twisting forces that exist in our culture* (and within us) that are constantly trying to pull us from a pattern of healthy eating. These forces include our own misperceptions about ourselves and what we eat (for me, no more 750-calorie muffins!). Refuse fast food. Don't get tricked into accepting big portion sizes. Don't let social pressures—friends, work situations, travel—derail healthy eating. And don't get bamboozled by the "healthy" or "organic" labels on packaged food. These are, by and large, marketing tools.
5. Do not view the above as a "diet" but as a permanent lifestyle change.

I know this prescription for healthy eating is likely to sound disappointingly conventional. It is a message that has been advocated by people such as Michael Pollan in his book *In Defense of Food* and Marion Nestle in *What to Eat*. I realize I haven't provided a revelation like "blueberries will make you smart and thin" or "bacon will melt your tummy fat." I wish I could. But those kinds of scientifically verifiable revelations do not exist. Simplicity *is* the revelation. Every expert I talked to, even those whose careers have been dedicated to nutrition research, agreed that the general parameters of a good diet are well known: fruits, vegetables, whole grains, nuts, lean meat or meat substitutes, and low-fat dairy. Any other advice is just chatter that twists the truth.

Yes, there are many additional details you can consider to make the pattern of eating even healthier, and I'll touch on some of them below. But few people come close to satisfying the very basic requirements of a nutritious diet. One study, published in 2009, found that fewer than one in ten Americans eat the recommended amount of fruits and vegetables. Another study, published in September 2010 and authored by a team from the National Cancer Institute and the National Center for Nutrition Policy and Promotion, paints an even grimmer picture. After examining the dietary habits of more than 16,000 people, the authors concluded that "nearly the entire US population consumes a diet that is not on par with recommendations." *The entire population*. Given this bleak reality, tinkering with details in order to optimize your diet is, for most of us, premature and absurd.

The three simple goals in my single-sentence summary—small portions, no junk, and lots of fruits and veggies—would have a radically beneficial impact on the diet of most North Americans. In fact, the discovery of these three simple dietary principles—more than halfway through my search for truth in relation to healthy living—have had the biggest effect on my life.

To confirm that my stripped-down approach was on track, I called on Marlene Schwartz, deputy director of the Rudd Center for Food Policy and Obesity at Yale University, who has spent years studying nutrition policy, especially in relation to kids.

Schwartz agreed that we need to keep diet advice simple. She said my daily plate approach, outlined above, seemed sensible.[8] "Eating 50 percent fruits and vegetables would be a huge improvement," she said. But she told me that we also need to recognize the reality that any room for any "discretionary calories" (candy, pop, chips, and other forms of empty calories) should be small. "Where governments fail is in the message of moderation. It's wrong. This is where the industry has had an influence. Industry is responsible for the 'everything in moderation' concept, the idea that there are no bad foods and no good foods so long as you eat everything in moderation."

Like my Food Advisory Team, Schwartz believes some foods are just plain bad—or, in the language of my diet plan, "poison foods." You cannot eat bad food *in moderation* because the food itself is so devoid of nutritional value. This is another industry-mediated pop-culture twist. If you are going to eat a nutritionally healthy diet and consume around 2000 calories a day, there is no room for a *moderate* amount of crap. In Schwartz's view, the daily plate only has room for a tiny slice—"perhaps 100–200 calories a day"—of crap. This 100–200-calorie allotment gives people the discretion to consume nutritionally vacant treats. But it isn't a lot (think *two* Oreos). "By putting this on the plate," she told me, "it shows how little you can have. It shows there isn't much room for extra stuff. Also, it shows that it is part of the same plate. You need to make a decision."

Schwartz's advice makes sense. And it fits with the research that suggests that complete restrictions do not work, especially for children. "I let my kids have one dessert a day," she reveals. "This includes chips, sugar cereal, French fries—all that stuff [counts as dessert]. I need to be careful because my kids get creative and there is a slide toward tremendously elaborate desserts—one cookie ... stacked with whipped cream and chocolate! And they rat each other out if there is cheating. But, in general, the approach works. The idea is to teach them balance."

"Dad, my hands are greasy," said my six-year-old son, Michael, as he finished a piece of bacon. "Can you open my Fruit Loops for me?" As I reached for the little box, I looked out the window and saw sea, mountains, islands. We were about halfway through our Alaskan cruise. We had boarded our ship, a luxurious monstrosity called the *Zuiderdam,* a few days earlier in Vancouver. The first thing the ship officials instructed us to do was head to the free buffet on the lido deck. The binge began before we left the dock. The supply of genuinely tasty food had been non-stop ever since. A tormenting experience. I felt like a recovering heroin addict hanging with Keith Richards during the recording of *Exile on Main Street.* So much hunger. And so much food just *wanting* to be eaten.

I had weighed myself moments before we left the house for the airport. My weight was 183 pounds. Conventional wisdom (meaning, what I found out on Google) has it that, on average, a person will gain eight pounds during a one-week cruise. Slightly more than a pound a day. That kind of weight gain would erase most of my progress, and so I embarked upon this voyage with a vow to stay true to my pattern of healthy eating.

"Dad, open my Fruit Loops," Michael repeated. I opened the box, and before I could stop myself my lower brainstem instructed me to stuff a few of the colourful sugar circles into my mouth.

"Dad, stop! Don't relapse!" It was my eldest son, Adam. He grabbed my hand and yanked it away. He looked into my eyes. "You're doing so well. Don't give up."

More than 10 million North Americans go on cruises each year. Our ship held approximately 2000 passengers. At eight pounds a day, this ship was on course to produce 16,000 pounds of human fat in one week. A floating fat factory. In order to avoid becoming part of this gluttonous gauge of holiday happiness, I followed a strict regime. I ate a healthy breakfast, minimal snacks, a small salad for lunch, and, as much as was humanly possible, a small dinner with no dessert. The final two stipulations were especially difficult to adhere to. When you're laughing, joking, and generally enjoying the company of loved ones—who are all eating elaborate and delicious-looking meals—it's all too easy to slip into a what-the-heck-I'm-on-holiday mentality. But, amazingly (and even as I write this, I am amazed), I found the willpower to stay on target. The cruise was enjoyable, but I might have been the only person on that boat for whom the meals were not a significant part of the journey.

Minutes after we arrived home I ran upstairs to the bathroom. I even left our suitcases in the car. I couldn't stand the suspense. I stripped naked and hopped on the scale: 183 pounds.

The same?! How could that be? I had exercised intensely every day of the cruise, usually for at least an hour. I had rarely ordered an entree, and even when I did, I ate only half of it. I avoided the freshly baked chocolate chip cookies. I drank almost no alcohol (five drinks the entire cruise). I had dessert only on the final night. And, on top of all that, I irritated my family with my annoying, self-righteous calorie counting. All that sacrifice and irritation and I had simply maintained my weight?

Lesson learned, yet again, from a different experiment. It was either a glass half-full (victory over excess), or a glass half-empty (another testament to the difficulty of weight control). You pick.

Before we leave the topic of diet, let's run through some additional nutritional facts and myths that emerged from my research. What follows is hardly a comprehensive list of new findings, but these topics came up again and again. And they were also issues that came up in discussions with friends and family.

Eat breakfast. This near-platitude, long championed by mothers throughout the world, has been confirmed by good research. In fact, this was one of the first principles Nick Wareham, the nutrition expert I talked to in Cambridge, brought up. "When you eat is important. Eat a big breakfast," he advised. Wareham has the research to back this up. His team examined the eating habits of almost 7000 individuals for several years. They found that "although all participants gained weight [again, that depressing reality of the inevitability of weight gain with age], the increased percentage of daily energy consumed at breakfast was associated with relatively lower weight gain." Other studies support this view. A 2009 study concluded that "being obese was associated with a meal pattern shifted to later in the day and significantly larger self-reported portions of main meals." So the rule should be, eat more in the A.M. than the P.M. Big breakfast, small dinner.

Don't eat in front of the TV. Research has shown that this habit leads to increased calorie intake. It's just automatic eating. There also seems to be an inverse correlation between watching TV and successful dieting. This is likely due, in part, to a general lifestyle and socio-economic status associated with watching TV. People who don't watch much TV probably have a healthier lifestyle. Less TV also means less exposure to food advertising, which can only be a good thing. (If we ate what was advertised on TV, one 2010 study found, we would eat 25 times the recommended amount of sugar and 20 times the amount of fat.)

What about snacking throughout the day? Over the past few years there has been much talk of the "five small meals" approach, the idea that eating throughout the day will help us avoid getting too hungry and bingeing. It has also been said that snacking keeps your energy levels more consistent. In fact, the data on snacking is inconclusive. Some studies support the idea of healthy snacking. Others associate it with weight gain. The most recent research, including a 2010 study published in the *American Journal of Clinical Nutrition,* found that snacking is inversely related to obesity. If you snack, you are less likely to be obese. However, *do not view snacking as an excuse to eat more.* Snacks should be healthy. Think fruits, vegetables, and nuts. (Note: many industry-produced and -processed "healthy snacks," such as cereal bars and muffins, are packed with more sugar and calories than a generous portion of ice cream, a lesson I learned the hard way.) And, don't forget, all snacks count toward your daily caloric intake.

It goes without saying (but I will say it anyway) that a late-night junk-food binge is a bad idea. Research has consistently shown that late-night eating is associated with weight gain. Do not get sucked in by the "fourth meal" marketing that many fast-food companies are now pushing. This is the late-night, after-party, post-clubbing meal we are being told is normal. It's another twisted message, this one pushing the idea that overeating is a fun way to cap a wild evening with your buddies. Some advertisements are subtle: thin, hip, and happy-looking young adults head for fries and a burger after a night on the town. Others are more direct. Taco Bell has an ad campaign called "Fourthmeal," which is "the meal between dinner and breakfast." The Taco Bell website hammers home the point: "You're out. You're hungry. You're doin' Fourthmeal." This is definitely not healthy snacking.

There has been a huge amount of research on the details of a healthy diet. Beyond the basic parameters outlined above—that is, eating fruits, vegetables, whole grains, and a bit of lean meat and dairy—researchers throughout the world have explored the particulars of the healthiest approach to nutrition. This work is not about radical new approaches to eating. No amazing thinning, energizing diets have been discovered. Rather, the research refines what a health-enhancing diet looks like. By adopting the basic prescription for a healthy diet I have described in this chapter, most of us will enjoy enormous health benefits, particularly since so many of us do not come close to meeting its requirements. But if you feel you have achieved the fundamentals, research suggests that more can be done. And what does this research tell us?

Eat like a Greek. Over the past decade, research has consistently shown the Mediterranean diet to be associated with decreased mortality and morbidity. People who eat like a Greek seem to lead longer and healthier lives. A systematic review of all the available evidence on the Mediterranean diet and its impact on mortality and on the incidence of chronic diseases was published in the *British Medical Journal* in 2008. The authors concluded that eating a "Mediterranean diet can significantly decrease the risk of overall mortality, mortality from cardiovascular diseases, incidence of or mortality from cancer, and incidence of Parkinson's disease and Alzheimer's disease." The overall reduction in mortality was significant, about 9 percent.

What is a Mediterranean diet? A 2009 study, also published in the *British Medical Journal,* examined the diets of more than 23,000 men and women. They found that the key components of the Mediterranean diet that predicted better health are moderate consumption of alcohol, low consumption of red meat, and high consumption of vegetables, fruits and nuts, olive oil, and legumes. They found only a minimal contribution for cereals and dairy products (meaning, these foods were not as important to good health). Another aspect of this diet, highlighted in a variety of other studies, is the weekly consumption of fish. This is tied, in part, to all the talk of the health benefits associated with omega-3 fatty acids, which have been shown to help with the prevention of cardiovascular disease and, perhaps, other ailments including dementia. But fish is also a healthier source of protein than, say, red meat, which is de-emphasized in the Mediterranean approach.

Emerging research is exploring the benefits of other dietary patterns, including the Paleolithic diet. This does not mean that you should eat what you kill. It means restricting your diet to the consumption of the foods that were available before the agricultural revolution (that is, 10,000 to 15,000 years ago). This is the food (so the thinking goes) we are evolutionarily designed to eat. But, on the whole, most of the experts I talked with felt that existing data supports a Mediterranean approach to tweaking the basics. For most of us, this simply means more fish (perhaps twice a week), nuts and legumes, and a reduced emphasis on dairy and grains.

You will have noted, no doubt, that the Mediterranean diet also includes the daily consumption of alcohol. And this provides me with an excuse to segue into a discussion of two of my favourite beverages: beer and coffee. Numerous studies have shown that the *moderate* consumption of alcohol lowers the risk of cardiovascular disease, especially for men. The data is less clear for women (there is some concern about a concomitant and offsetting increase in cancer risk). But, if you are a man, having one or two drinks a day is probably a good idea. (My Food Advisory Team agreed with this conclusion, but also noted the calorie challenge. Two bottles of beer, for example, is about 300 calories. Beyond the heart-protective effect of the alcohol, beer—and, for that matter, other forms of alcohol—is not a nutrient-dense diet option. For my diet plan, I have opted for half a bottle of beer a couple of times a week. Pathetic, I know, but my days of hard partying are behind me anyway.)

Coffee is even better for you. Given that I spend a significant portion of my life (and salary) in cafés that sell overpriced espresso products, the continued accumulation of data promoting the health benefits of coffee has been personally gratifying and guilt-reducing. I may not be able to afford to send my children to university, but at least the family fortune was not squandered on an unhealthy addiction. Studies have found that coffee (no sugar or cream, naturally) may protect against some cancers and decrease the risk of heart disease, diabetes, and Parkinson's. It also seems to boost cognitive function and athletic performance. While the data is still coming in, I think it's fair to say that, at a minimum, coffee is not bad for you and probably even healthy. Tea can have similar effects, but I don't like tea, so I didn't spend nearly as much time researching it as I did coffee.

As with fitness, there are a number of diet inaccuracies and half-truths floating around the pop-culture universe. I'd like to note the three most frequently mentioned by the experts. These myths were also viewed as the most worrisome, given that they are often associated with unfounded and potentially harmful diet products: first, the idea that we need to detoxify or cleanse our bodies (I'll return to this bogus concept in the remedies chapter); second, the notion that we need a plethora of dietary supplements; and third, the idea that eating a specific food or product will increase our metabolism so that we can lose weight. There is no evidence to support the benefits of detoxification, and the research on supplements is complicated but, in general, leans toward avoidance.[9]

So, if you see the words *detoxify, cleanse, supplement,* or *metabolism* associated with a product or process, be suspicious. Someone is trying to sell you something that likely does not work and, in fact, might be harmful.

But let's get back to bacon. Is healthy bacon fact or fiction? Can my family gorge on its greasy goodness? Alas, I could not find anything to recommend the consumption of large amounts of bacon. The sodium and nitrates found in this crispy meat product are far from good. Despite reports to the contrary, bacon remains a poor, albeit delicious, food choice.

With just nine days left until my "after" body fat calculation with the DXA machine, I took stock and made some notes to myself. Three months of healthy eating. Three months of calorie restriction. Three months of no poisons. I felt great, but I was anxious to find out if any of this sacrifice had actually reduced my body fat. I confess that I was hoping to slide into the "fit" category, but I also worried that I might have lost an equal measure of muscle. The last week of the diet felt like the sprint at the end of a marathon. It took considerable effort during that final week to stay away from cookies and chocolate ("I'm basically done, right? It won't matter now, right?").

There was one last meeting with FAT. We convened at a local

café, where we consumed exactly half a glass of beer each. Three cheap, but classy, dates. I ran through my major conclusions for them, and the FAT agreed that I'd hit all the right notes.

Not only that, they each commented upon what was now apparently my noticeable weight loss. Kim Raine, fit and trim herself, joked that she needed to start following her own advice. I was a full 20 pounds lighter than when I'd started my experiment. Twenty-three pounds, to be exact. I was now approximately the same weight as when I finished my undergraduate degree, 23 years ago. A pound a year. This weight loss was not achieved via exercise intensity (since I've exercised intensely my whole life, anyway). It was not achieved through long walks, runs, or bike rides (since I'd also done my share of endurance training and had never really lost any weight). It was not achieved by following some weird and wonderful diet (I ignored all "diet food" advice, and had no focus on carbs, fat, or protein). It was not achieved by taking a diet pharmaceutical pill or supplement (which studies have shown don't work any better than placebos).

I achieved these results by following a straightforward and healthy eating plan. And more than anything, I simply ate fewer calories.

As I left the café I bent down to open my bike lock, and my jeans fell halfway down my ass, offering up a classic plumber's crack for the café's patio customers. I was certain the image had an immediate appetite-suppressing effect on everyone unfortunate enough to catch a glimpse. I laughed to myself, but then another thought crossed my mind. Would I need to buy new clothes? Or was I destined to return to my old ways at the end of the experiment, and put all the pounds back on?

This simple question led me directly to one last, depressing twist.

"Calories are the currency of weight loss," Ayra Sharma told me, using an of-course-everyone-should-know-this tone. Sharma is a professor of medicine at the University of Alberta, and the scientific director for the Canadian Obesity Network. He has spent his career exploring both the health issues associated with obesity and the various strategies used to lose weight. His calorie currency comment was made in response to my bragging: I had just told him, perhaps with a tinge of smugness, about my successful weight-loss efforts. He was unimpressed.

"You have simply eaten fewer calories for a few months. Good for you. But when you put the weight back on, which is going to happen, it will just get more and more difficult to take it off again. And the weight," he added after a short pause, "comes back quickly."

Though I was well aware of the dismal data on long-term weight loss, Sharma's dose of realism was still demoralizing. But a sober reading of the existing diet research incontestably supports his conclusion. And this reality, so twisted by the diet industry, is the depressing note that I feel compelled to relate. People can't keep the weight off. Well, the vast majority of us can't anyway, no matter what diet we use or how often we diet. It does not really matter if you go low-fat or low-carb. You can lose it, but it's tough to keep it off.

A 2005 analysis of commercial diet programs, for example, concluded that "about 90 percent of people who diet gain every pound back that they lose regardless of their weight-loss method" and that there was little good evidence "to support the use of the major commercial and self-help weight loss programs." Other studies support this view, including two that are often referenced: a 2005 report in the *Journal of the American Medical Association (JAMA)* and another in the *New England Journal of Medicine* in 2008. The studies systematically examined a variety of diet programs in a clinical setting, programs such as the Atkins (low-carbohydrate) diet, Weight Watchers, the Ornish diet (low fat), the Zone, and the Mediterranean diet. It was found that all these diets could result in some weight loss beneficial to health. But in each case actual, sustained weight loss was, at best, modest.

The study published in *JAMA*, for example, found that weight loss was similar across all diets and that about 25 percent of the participants lost 5 percent of their initial body weight in one year (so, if you are a 200-pound male, that's about ten pounds). But, again, long-term adherence was dismal. Over a one-year period, there was a steady increase in self-reported caloric intake, regardless of diet. The diets slowly fell apart.

So, even when the diets are a planned part of a research project and involve health-care professionals, regular counselling, and motivated and highly educated subjects, weight-loss goals are rarely achieved or sustained. There is some slight variation in the performance of specific diets. Some research suggests, for example, that diets that minimize carbohydrate intake and emphasize energy-dense foods (such as the Atkins) are more effective because dieters feel more satiated. But a 2009 study of more than 800 dieters, published in the *New England Journal of Medicine,* found that all diets had similar, relatively modest outcomes. A 2010 study of over 300 dieters came to a similar conclusion. The key to keeping it off is simply to eat less in perpetuity, which is a feat few of us can achieve. As one writer commented in the *New England Journal of Medicine* in 2009, "Evidently, individual treatment is powerless against an environment that offers so many high-calorie foods ... It is obvious by now that weight losses among participants in diet trials will at best average 3 to 4 kg after 2 to 4 years and that they will be less among people who are poor or uneducated, groups that are hit hardest by obesity."

Ayra Sharma's cutting comment about my inevitable post-diet weight gain—"like an elastic band snapping back to its resting state" was his colourful, if unwelcome, metaphor—is supported by ample empirical data. A 2003 study of over 2000 individuals who had successfully lost weight found that after gaining back a few pounds, only 4.7 percent were able to get back again to their lowest weight and only 12 percent re-lost at least half. Again, pretty grim statistics, especially when you consider that this was a study of individuals who possessed the willpower to lose weight in the first place.

When you see a picture of a trim and beautiful individual on the cover of a weight-loss product—whether it's a book, video, food, or pill—ask yourself if you are a few pounds away from achieving that look. Because that is, on average, the best you can hope for, long term, by embarking on a diet. A few pounds less than you are now.

Even liposuction doesn't work. Even if you have a machine suck the fat out of your body, that fat will likely come back—albeit in a different spot. A 2011 study published in the journal *Obesity* found that individuals who had had liposuction on, say, the thighs put the weight back on in the tummy. As noted throughout this book, we humans are depressingly efficient calorie consumption machines. We seek status quo. No matter how the pounds are taken off, our bodies find a way to creep back to the previous weight.

The inevitability of failure is, ironically, what drives the diet industry. This may seem counterintuitive: you would think that the failure to deliver as promised would kill the business, but the opposite is true. "The diet industry is the best business to be in," says Sharma. "It's easy to deliver some short-term weight loss. If you put weight back on—and everyone does—the industry can say it's your fault. You didn't work hard enough. If you keep it off, they can take credit. But you won't keep it off, and the diet industry can say that it's your fault. You did something wrong. So you go back to the industry for more help. An endless stream of repeat customers."

One 2007 survey of dieters found that more than 30 percent of respondents had been on three to five diets in their lifetime, and a further 25 percent had attempted dieting at least 20 times. That is a pretty reliable customer base. And this perpetual demand has led to the creation of an immense business, one that brings in $60 billion a year in the United States alone.

The twist the industry is selling is the idea that there is a solution—a pill, a diet, or a special food—that will change your body forever. This solution does not exist. People have been dieting for decades, centuries. If something worked, we would know. "There will never be a diet, or a pill, or anything that will allow you to go back to what you did before," says Sharma. "There is no cure for weight gain. You must simply cut calories and stick with a strict diet ... forever. Ninety-five percent of people will put the weight back on. Only one in twenty can keep it off for any length of time. You must dedicate your life to it. It's a lifestyle."[10]

Diane Finegood agrees with Sharma. She's in a position to know. Ten years ago, when she became the scientific director of a national research institute that hosts nutrition research, she lost 75 pounds. And she has kept it off. She is one of the 5 percent, an individual who has dedicated herself to weight maintenance. "I can only eat 1600 to 1700 calories a day just to maintain my current weight," she told me. "It's a constant battle."

Think about that: 1700 calories a day. That's not a lot of food, but it's all she's allowed herself to eat per day for the rest of her life. And given Finegood's lifestyle, there is constant temptation to eat more. As a leading academic, she ends up at a lot of receptions and work dinners. "I can't allow something to be put in front of me or I'll eat it. I only buy starters." (Sharma told me the same thing. He claims he hasn't ordered an entree for years.)

"You can't view [weight loss] as a diet. Don't think diet," Finegood said. "Think change in behaviour. Think about little things you can change and change for the rest of your life. Sustainable behaviour changes."

The diet industry, of course, is not built around this kind of long-term lifestyle change. It's built around the promise of looking good and of immediate results. Finegood emphasized this: "The problem with the diet industry is they don't sell health. They prey on people's desire to look a certain way—creating noise and promises that are not obtainable. But even if you don't lose much weight, eating nutritiously will result in physical change that is good for you."

As with exercise, there are many reasons to eat healthily that are not related to aesthetics. I did not receive many spontaneous comments about my weight—meaning, I didn't receive many comments about how I *looked*—until I hit the 20-pound mark. It's entirely possible that this had more to do with my personality than my appearance, but I was unable to measure that. Still, there were many things I noticed: I felt a lot better. I had more energy. My only real health concern, my blood pressure, improved. (As we know, eating better, in particular by eating more fruits and vegetables, is one of the surest ways to fend off chronic diseases such as cancer and heart disease.) Yet despite all these benefits, weight loss for the purpose of looks was the *sole* thing people wanted to know about when they heard about my diet. I did not get a single inquiry from friends, family, or colleagues about the health benefits. Not one. Everyone wanted to know how to lose weight and keep it off.

People just want to look good, I guess.

As soon as I left the DXA lab I phoned Joanne with my results. I was standing beside the bike rack next to the building that houses the nutrition laboratory. I knew she was as curious as I was. I read Joanne the exact results. My numbers for the first test: 193.8 pounds and 18.7 percent body fat. My "after" results: 175.1 pounds and 10.2 percent body fat. And I'd lost virtually no muscle mass.

I was staggered by the outcome, as were the lab technician ("I don't think I have ever seen numbers like this"), my FAT ("Don't lose any more!" Linda warned), and Joanne ("I guess not eating a dozen cookies every night pays off"). I had already lost four or five pounds before the first DXA, which meant that my total weight loss was almost 25 pounds. And my body fat now was as low as a middle-aged man should go.

All due to simple eating. Smaller portions. No poison. Healthier choices.

Feeling both gratified and, admittedly, a touch self-satisfied, I left the lab and biked back toward my side of the campus. I passed a convenience store. I zipped in and got a bag of M&Ms. Hey, I told myself, I've earned it. And after all, how much damage could one little itty-bitty bag do?

3

GENETICS
BLUEPRINT FOR A HEALTHY LIFE?

On a recent flight home from Europe the movie selection was so limited that I decided, reluctantly, to watch *The Time Traveler's Wife*. My reluctance wasn't because of movie snobbery (I have lowbrow taste), but because when I'm tired and 30,000 feet above the ocean, manipulative, sentimental romances make me cry, and not just sniffles, but full tears occasionally accompanied by sobbing. Only it didn't happen with *The Time Traveler's Wife*. I soon found myself too preoccupied with the premise to let loose with sobs.

The protagonist, played by Eric Bana, has a genetic anomaly—chrono-displacement disorder, to be exact—which causes him to involuntarily travel through time (and which, it turns out, is tough on a relationship). While I am fully able to suspend disbelief in the service of cheap entertainment, I found it bizarre that the creative minds behind the movie—and, I assume, the book—felt that genetic science was the best way to explain time travel. It struck me as the equivalent of using cellular biology to describe gravity.

The more I thought about it, however, the more Bana's chromosomal dilemma made perfect sense. It echoes a popular theme: these days genetics is all-powerful, it's everywhere, and we can't escape its reach. It's only natural that Hollywood would embrace genetics as a mystical force powerful enough to tear Bana from the embrace of the beautiful Rachel McAdams.

For decades we have been bombarded with headlines and magazine covers that extol advances in genetics. We have been told that we are living through a "genetic revolution." In 1994, *Time* magazine ran a cover story entitled "Genetics: The Future Is Now." A decade later, *Time* ran another genetics cover story, telling us this time that "Gene Science Has Changed Our Lives." If you believe the headlines, scientists have already located a gene for virtually every human condition you can think of. Even a short sampling of recent headlines bears this out: "Is 'Laziness Gene' to Blame for Couch Potatoes?"; "The God Gene: Does Our DNA Compel Us to Seek a Higher Power?"; "Always Lost? It May Be in Your Genes"; "Party Animal: It May Be in Your Genes"; "Marriage Problems? Husband's Genes May Be the Problem"; and my personal favourite, "Genes May Affect Popularity, Researchers Say."

If there is any area of science research that seems likely to help us live a healthier life, surely it would be genetics. Genes represent the cards we have been dealt. They form the immutable biological groundwork on which all other health issues are built—or so we are told. Consequently, an exploration of genetics seems the logical place to go next in my quest for the truth about what makes us healthy. We have seen that while no miracle diets or exercise routines exist, a relatively straightforward approach to fitness and diet can do much for our health. Perhaps an exploration of our genes can offer more.

But what is the truth? While I am fairly certain my genes can't transport me to another time or dimension, the extraordinary level of pop-culture attention devoted to genetics certainly gives the impression that a revolution is afoot. If my genes can tell me whether I am destined to be lost, religious, lazy, unfaithful, or unpopular, and if, as the magazine covers declare, genetics' future is now, then shouldn't I be able to put genetics to work for me?

The first time I was in Mountain View, California, was in 1997. I was at Stanford (which is next door in Palo Alto) as part of a research team looking at the social implications of the "genetic revolution." Should kids get tested to find out their genetic predispositions? Will insurance companies want to know your genetic background?

These questions seemed timely and important. At that time we were still only a few years into the so-called "revolution." The massive Human Genome Project was not yet finished and it was easy to believe that society was only a few years away from enjoying myriad medical breakthroughs and concomitant social dilemmas. Few of us within the biomedical community doubted the potential of the field. I toiled away producing articles on how society should respond to the imminent revolutionary breakthroughs.

Now, in 2010, I was back in Mountain View to take advantage of one of the few tangible products of that revolution. I was there to secure a profile of my genes—or, at least, almost 600,000 genetic markers—by a company called 23andMe.

My journey to 23andMe actually started in an auditorium at the University of Toronto in the fall of 2008. The event was a public lecture on direct-to-consumer genetic testing, and the auditorium was surprisingly full with a crowd I estimated at well over 300, from all walks of life. I was sitting on the stage with Joanna Mountain, a Stanford professor and the senior director of research for 23andMe. The focus of the event was whether the new era of commercial genetic testing offered any health benefits. Could genetic testing really provide the average person with information to improve their health?

Mountain later told me that that Toronto panel, which included a number of scientists and genetic clinicians from around the world, "was one of the most hostile panels I have ever been part of. I felt like it was me against six other people."

I was one of the six.

When it was my turn to speak I told the audience that most of the information provided by genetic-testing companies was useless. A complete waste of money. In response, in front of everyone, Mountain calmly offered me the 23andMe service for free. It sounded like a dare.

Onstage, I met Mountain's strategic generosity with a firm and public "no thanks." I declined for three reasons. First, I truly believed the information would not tell me anything particularly valuable about my current and future health status. Second, despite my skepticism, I was worried that I was just neurotic enough to overinterpret whatever results were provided. In other words, I'd become one of the "worried well," a perfectly healthy individual who is constantly concerned that a slight genetic predisposition will blossom into some horrid disease. And third, I didn't want to come off as a hypocritical opportunist. ("Well, if you're offering it for free, sign me up!") I only let the audience in on the first reason.

But my curiosity grew. Perhaps I *was* interested in knowing this stuff. I am, after all, a self-interested, vain, and superficial person: more information about *me* might be fun. So, more than a year after the lecture, I took Mountain up on her offer. She responded immediately, sending me the information necessary to order a 23andMe test kit from the company website.

When I told my family of my plan, most were not thrilled. My wife was worried I'd discover health information that would stress her out. My eldest daughter, eleven, wondered why "anyone would want to know what will kill them." And my six-year-old son viewed the required spit-collection process—23andMe sends you a tube to gather what seems like quite a bit of spit (a true YouTube, I suppose)—as a significant and insurmountable problem. "That is super gross, Dad. Do it in the bathroom."

I thought I should also run the idea past my two biological brothers, just to see if they had any concerns about the unveiling of predisposition information. Whatever I found out about myself might have relevance to them. "I don't care about that privacy shit," was one reaction. "Go ahead and get tested." My other brother was slightly more apprehensive, but he's an artist and therefore thrives on apprehension. He gave me the okay, too.

With these lukewarm family endorsements secured, I sent a large tube of lukewarm saliva across the Canada–US border.

Why get your genes tested? Trying to answer this question requires a look at the scientific advances that have brought us to the point where a comparatively small company such as 23andMe can test more than half a million genetic markers for less than the cost of a cheap suit. Genes, as anyone with even a basic exposure to biology knows, are the units of heredity—a reality first illuminated in the 1860s by Gregor Mendel, an Augustinian monk who diligently and meticulously studied the variation of pea plants in the courtyard of an abbey in what is now the Czech Republic. Mendel showed that traits, such as wrinkly pea skin, could be passed from one generation of pea to another.

In the early 1900s, scientists built on and refined Mendel's rules of heredity, including its role in the development of disease. In 1908, for example, British physician Archibald Garrod suggested that some diseases were due to "inborn errors of the metabolism." It wasn't until 1953 that the structure of the unit of heredity, DNA, was famously described by James Watson and Francis Crick in a ridiculously short article in the journal *Nature*. In their one-page paper, one of the most famous in science, they start with the understated pronouncement that "they wish to suggest a structure" for DNA that has "novel features which are of considerable biological interest." They then go on to describe the double helix shape that has become the ubiquitous emblem of modern science.

In the 1970s and 80s a variety of new technologies allowed scientists to read the chemical codes that make up our DNA. The complexity is staggering. Try to imagine your genetic material, your *genome*, as a string of letters ... three *billion* letters. It would take ten years to read it out loud. New technology (sequencing) made it possible to start unravelling this string and to try to identify the place and function of genes. While slow and expensive compared with today's computer-assisted sequencing, the 1980s technology progressed to the point where many in the scientific community thought it was time for a big step: the sequencing, or mapping, of the entire human genome. In April 1987 a group of leading scientists issued a report calling for the US government to "fund a major new initiative whose goal is to provide the methods and tools which will lead to an understanding of the human genome."

This report was what initiated the Human Genome Project (HGP), the project that led to the *genetic revolution,* the *era of genetic medicine,* the *biotech century,* or whatever other hyperbole-laden label you wish to use. The authors of the report were enthusiastic about the potential of the project, understandably, since they were asking the US government for hundreds of millions of dollars, but they did add a caveat, tipping their hand to the unknowable meandering of scientific inquiry: "We do not know what sequence information is the most valuable. It is likely that the most significant applications to medicine cannot be foreseen at the present time."

The Human Genome Project was biology's first foray into the realm of "big science." It cost more than $3 billion, making it a scientific effort on the scale of the Manhattan Project and the trip to the moon. Measured against quantifiable technical goals, it was a genuine success. In April 2003, 99 percent of the gene containing part of human sequence was finished to 99.99 percent accuracy. The project was, in effect, complete.

Francis Collins, the head of the US arm of the HGP and currently the director of the National Institutes of Health, has consistently said that improving human health and reducing the burden of disease for all people was, and remains, the real goal of the HGP. Measured against this standard, the success of the HGP is less certain. In 2001, Collins authored a paper with Monique K. Mansoura in which they stated that "the most critical measure of the success of the HGP will be determined by the answer to this question: To what extent did the scientific and medical advances derived from the HGP reduce the burden of disease for all people?"

Since the completion of the HGP the technological advances have been breathtaking. Gene sequencing can be performed at a speed unimaginable even a few years ago. It is also vastly cheaper. A 2010 article in the journal *Science* outlined how a genome was sequenced for just $4400. Many in the scientific community think that cheap genomes (as little as $500) are just around the corner. In only seven years (a true blink of the eye in the context of the usual tempo of science), the cost of sequencing a genome has dropped from the billions to the thousands, and soon, it seems inevitable, to the hundreds. It's like the Apollo moon project now costing the same as a flight from Edmonton to Toronto.

These are the technological advances, driven by computer automation, that have allowed companies like 23andMe to establish a viable business. The cost, speed, and efficiency with which genetic markers can be analyzed (23andMe does not sequence entire genomes; rather, the company looks at genetic markers along the genome, which, as a result of genetic research, are known to be associated with a disease or inherited trait) have reached the point where the analysis can be offered to the public at a reasonable rate.

But what does all this mean to me, for my health? Yes, it's exciting, but can this technological genetic wizardry allow me to live a healthier life? I had sent my saliva south. I was now waiting for answers, wondering if my 23andMe genetic profile would provide me with information I could use to make health decisions that matter.

Two days before my scheduled meeting with Joanna Mountain in California, I received an email from 23andMe telling me that my results were ready. The email said I could even look at the results now, on my computer. I was sitting in my home office when I got the message, enjoying a guilty few minutes surfing the net at precisely the hour I knew two kids were headed to bed and two others needed help with their homework. Perhaps not the best time to check into one's genetic destiny, but I was too nervous and excited to resist. I opened the company website. I found my profile. There it was. I could hear the normal family chaos rumbling outside my office, but I opened the results anyway. They were intriguing. Provocative. But then I read something …

"What the fuck?" I said, obviously louder than I'd meant to. I don't often use four-letter words. Even from within my office, the outburst caused a brief and instantaneous moment of family silence.

I arrived at the 23andMe headquarters early in the morning. It was a typical business-park office. The bland brick facade and modest corporate sign conveyed "insurance adjusters" more than "paradigm-shifting hub of innovation." Somehow one expected a more assertive presence. This company, after all, has garnered considerable international attention and provoked a good deal of controversy. In 2009, *Time* magazine declared its testing kit, the one I used to collect my spit, to be the innovation of the year. Meanwhile, a number of jurisdictions, including California and New York, have locked horns with 23andMe, claiming, in brief, that they are providing an unregulated medical service. None of this was evident from the parking lot.

The building looked dead. The door was locked. I pressed a button, and soon a hip-looking, youngish employee opened the door without even asking who I was. Clearly theirs was a relaxed working environment. Joanna Mountain arrived a few moments later. "Sorry I'm late," she said apologetically. "Dropping the kids off."

Mountain, whose look and speech are more professorial than corporate, walked me through the open-plan office. We sat in a brightly coloured lunch room adjacent to an even more brightly coloured workout room. "We have a Google mentality about fitness and lifestyle," Mountain explained, referencing one of 23andMe's biggest investors and Mountain View corporate neighbour.

We logged onto my personal 23andMe webpage and started chatting about the results. These were beautifully arranged on the screen. Each "disease risk" is presented with a "confidence" rating (based on available research) and a listing of your risk, expressed as a percentage, compared with the general population. The website offers results for dozens of diseases. They are grouped together according to *elevated risk, low risk,* and *typical risk.*

Here is what I found out: I do not have an increased risk for throat cancer (the disease that killed my mother); I do not have an increased risk for high blood pressure (despite the fact that regular blood pressure checks tell me my systolic pressure is consistently elevated); and I have an increased risk for celiac disease, prostate cancer, and atrial fibrillation.

At first glance, these findings all appear somewhat ominous. Being told you have an "increased risk for prostate cancer" is not a great way to start the day. But were these risk assessments meaningful? Not really. Almost without exception, the variance from the average was relatively slight. And more importantly, they must be read within the context of daily life.

For example, my 23andMe report tells me that my genetic markers indicate I have a 0.4 percent chance of getting celiac disease. This is an increase over the population average of 0.1 percent. I also have an increased risk of multiple sclerosis. My risk: 0.5 percent. The population average: 0.3 percent. When compared with the impact of obesity, not exercising, driving while talking on a cellphone, and, the big one, smoking, a 0.2 percent increase in risk is essentially lost in the wash of the risks associated with life. Meaningless. And these two risks are highlighted in the 23andMe report as among my most significant genetic risks.

There was still the one piece of genetic information that had so unnerved me when sitting in my home office. I asked Mountain about that one piece of data. "Can that genetic result be right?"

"Probably," she responded.

My expression must have betrayed my disappointment because she added, quickly and with a patient smile, "Perhaps you have another genetic mutation that moderates its impact."

In a 1999 *New England Journal of Medicine* article, Francis Collins suggested that the main benefits of the Human Genome Project are going to be new drug therapies, gene therapy, pharmacogenomics (that is, drugs tailor-made to match an individual's genetic predispositions), and preventive medicine (using genetic risk information to motivate behaviour change). Collins used a hypothetical scenario to describe how, in 2010, the genetic revolution would impact our lives and the practice of medicine. The scenario represented a hopeful prediction made while the excitement surrounding the Human Genome Project, and the future of health care, was at its most intense.

Collins's predictive scenario involved a visit by a typical male patient to a doctor in which the doctor uses an interactive computer program to take the patient's family history and in which the patient takes a battery of genetic tests for common diseases (again, using an interactive computer program to pick the tests). The history and tests indicate that the patient is at high risk for lung and coronary disease. And, as Collins put it in the article, "Confronted with the reality of his own genetic data, he arrives at that crucial 'teachable moment' when a lifelong change in health-related behavior, focused on reducing specific risks, is possible." The genetic risk information has provided the patient with the "key motivation for him to join a support group of persons at genetically high risk for serious complications of smoking, and he successfully kicks the habit."

Well, 2010 has come and gone as I write these words and the genetic revolution has not manifested itself in the way Collins envisaged. Family physicians do not offer routine testing for genetic predispositions to common diseases, they do not have cool interactive computer programs to determine which genetic tests to order, and the seemingly unalterable tendency of humans to *not* alter a bad habit, no matter how harmful, remains intact.

To be fair to Collins, almost every prediction about the practical application of genetics has been wrong, or at least overly optimistic. In the mid- to late-1990s, the emphasis was largely on gene therapy. If you are old enough, you might remember the rhetoric that was bandied about at the time. For example, gene therapy was part of a January 1999 *Time* magazine cover story on the future of medicine. The article, entitled "Fixing the Genes," quoted a researcher saying that in the future, "virtually every disease will have gene therapy as one of its treatments." The idea was that doctors would fix the abnormal genes that cause disease, such as cystic fibrosis, by inserting a normal gene into the body to take its place. Normal function would be restored and the disease cured or, at least, the patient's health improved. An intriguing and worthwhile goal, no doubt.

However, hundreds of millions of dollars have been spent (post-2005, the US National Institutes of Health invested approximately $400 million) and there have been hundreds of Phase I clinical trials (these are small studies used to test safety), but not a single gene therapy has been brought to market or approved for general use by the relevant regulatory agencies. There have been a few successes, but only for rare diseases and only within the bounds of controlled research initiatives. In 2002, for example, researchers in Milan demonstrated the potential therapeutic benefit of gene therapy by treating a rare and deadly immune deficiency disease called ADA SCID. But, as a widely applicable treatment strategy, gene therapy has been a bust. This is largely because it has proved extremely difficult to get normal genes safely into a human body and to make them stay and function normally over an extended period of time.

To make matters worse for this troubled field, it is also associated with one of the most infamous research ethics controversies of recent history: the 1999 death of a gene therapy research participant, Jesse Gelsinger. The case involved allegations that the investigators were financially conflicted, that the ethics review was inadequate, and that pre-clinical data was insufficient to justify research on a human subject. Now, rightly or not, the Gelsinger case (a lawsuit was settled in 2005) stands as a cautionary tale of what happens when, pushed by public interest, or by economic and scientific pressure, a technology is rushed prematurely to the clinic.

As gene therapy's star began to fade, the emphasis in both the scientific literature and in the popular press swung to genetic diagnosis. The belief was that we would find many highly predictive genetic tests. This belief was reinforced by the 1994 (BRCA 1) and 1995 (BRCA 2) discoveries of genetic mutations (present in less than 1 percent of the general population) that increased the chance (for some individuals, dramatically) of developing breast or ovarian cancer. But scientists have yet to find the hoped-for numerous and highly predictive genetic mutations for common, complex diseases. In other words, with the exception of rare diseases that are directly caused by a genetic mutation—such as Huntington's disease or cystic fibrosis—we do not have a battery of genetic tests that can tell us, with any degree of confidence, if you will get a particular kind of cancer, heart disease, or other common health condition.

A few days after I got back from Mountain View I called Jonathan Kimmelman, a professor at McGill University. With a PhD from Yale in molecular biophysics and biochemistry and a membership in McGill's Biomedical Ethics Unit, he is a scholar with a strong understanding of both the science of genetics and its potential social consequences.

It would be fair to say that the hype surrounding the potential clinical application of genetic technologies irritates the usually mild-mannered Kimmelman. He becomes quite agitated when discussing the topic. He believes that researchers are far too confident in anticipating how basic research—conducted in labs using test tubes, computers, and, when necessary, animals—will eventually play out on a human subject. He points out that promising results from basic and pre-clinical research (done to explore such things as foundational principles and mechanisms of action) often do not translate into a successful clinical trial (research involving actual humans). "If we went back and looked at all the pre-clinical research that looked good and exciting, we'd see that 99 out of 100 times we were wrong. I marvel at how people interpret these pre-clinical studies. You'd think the reality of the situation would change our perspective. The way people interpret pre-clinical evidence is often shockingly unsophisticated."

Kimmelman also thinks that despite the glowing reports you see in the popular press, and by people like Frances Collins, there has, in fact, been a rapid deterioration in scientific support for the idea that genetic testing will provide vast amounts of clinically useful information. "In the mid-90s everyone was lined up behind the belief in the existence of highly predictive genes," he told me. "Around 2002 or 2003 that belief started to break up. People began to doubt its value. You still hear it discussed and presented as an exciting possibility, but no one really seems to believe it."

Kimmelman is something of an emerging star on social issues associated with biomedical innovation, but he is not directly engaged in genetic research. To find out if a leading geneticist would agree with Kimmelman's pessimistic vision, I went to the top of the Canadian genetic research heap. Few people in Canada have had as great an impact on genetic research, both scientifically and politically, as Tom Hudson. He has been the lead author on many renowned studies and was, for years, the director of McGill's genomic research centre. He's been both a participant in the genetic revolution (when the HGP money started to flow in 1991 he was at MIT, one of the major gene-sequencing centres) and an energetic advocate of genetic research. In the late 1990s, he was knocking on doors in Ottawa asking politicians for funding to establish Genome Canada, an entity that has since become the focal point for much of Canada's leading genetic research.

When I met with Hudson, his time was, characteristically, short. "How much time do I have?" I asked.

He glanced at his watch. "I can give you 25 minutes, maybe 30."

I have known Tom Hudson for more than a decade. We have worked together on policy committees, shared bad coffee at too many conferences to count, exchanged ideas for our respective research projects, and swapped stories about our children. Yet despite our long association, Hudson is so much in demand that I needed to make an appointment months in advance. For 30 minutes.

With time at a premium, I dove in. What, I asked him, is the value of genetic testing, right now, for my personal health?

"The genetic revolution will be more about the details of clinical care than the man on the street," Hudson said, leaning toward me across a table covered in science journals. "It will not help individuals make day-to-day decisions."

He answered so quickly, and with such confidence and intensity, that I was left with the impression that this was a statement he rolled out frequently. I was a bit taken aback by his bluntness, and so I asked him to expand on his answer.

"The science is moving so fast." He paused and exhaled before continuing. "But you need to find things that allow for prevention, detection, diagnosis, something with clinical value. Those things take time. You just can't find markers to predict disease. You need to acknowledge all the complicating factors, the environment, human behaviour ... everything. And if we find a genetic marker that predisposes someone to disease, we need to be able to modify the risks in a meaningful way." He leaned toward me again. "For example, maybe your genetics put you at a higher risk for a certain cancer. We can get those people to get more screening. But we still need to do the clinical research to find out if this screening really makes a difference. Is it worth it?"

He was alluding to the fact that there are many examples of screening procedures that have not panned out as originally envisioned. The prostate specific antigen (PSA) test is a good example. In March 2010, after it had been in common use for more than a decade, the American Cancer Society urged caution in the use of PSA testing as a screening method for prostate cancer. The decision was motivated by several studies that questioned the utility of the test and by a growing concern that testing was leading to unnecessary and potentially harmful investigations. For some, PSA screening was doing more harm than good. Hudson was cautioning that this scenario could play out with testing for cancer genes. Even if we find genes that increase an individual's risk of getting cancer, we don't know for sure if using the genetic test would be helpful or harmful.

I asked Hudson what he thought was driving the broad public misperception of the immediate or near-future clinical utility of genetic testing. Here, Hudson was in complete agreement with Jonathan Kimmelman. Too many basic researchers (and, as a result, many in the media) simply do not understand that good science does not necessarily lead to useful clinical applications. "Just because it's valuable science," he said, "doesn't mean it's clinically valuable."

There are many factors that add complexity to the transition from basic science to clinical application: the logistics of doing clinical research (as Hudson noted, "You need lots of time and big research teams"); gender and socio-economic factors (men typically don't like to get screened, for example); the lack of Big Pharma–style funding for clinical trials. The significance and difficulty involved in clinical research are undervalued, underfunded, and misunderstood, Hudson noted. "There are many great basic scientists doing amazing things, but they're clueless about how to bring something to the clinic."

It was becoming clearer and clearer to me that the relationship between genetic research and our physical health is considerably more complicated than we are often led to believe. The so-called "genetic revolution" is more of an uncertain and slow evolution. We have vast amounts of genetic information at our disposal, and incomprehensibly sophisticated technologies that allow for the speedy production of more and more of it, but we still appear to be nibbling indecisively around the edges of what to do with all the data.

As Hudson walked me to the elevator he continued to supply me with data about genetics, cancer, the speed of his sequencers, and, most tellingly, his agency's tobacco control work. "This tobacco project is having an impact, right now," he said, holding a door open for me. "But it doesn't have anything to do with genetics."

It seemed a striking thing for a world-renowned geneticist to say.

Given the underwhelming performance of gene therapy and predictive testing, much of the current genetics rhetoric is focused on using genetic information to encourage people to adopt a healthy lifestyle. The watchword is prevention. The plan: if we first obtain our individual genetic information, we can use it to make decisions about what to eat, what kind of exercise to take up, and what kind of regular surveillance testing to submit to. If our genes show that we're slightly more likely than our neighbours to contract a particular disease, might we be motivated to take evasive action? ("I'm at increased risk for heart disease. I had better start exercising and eating right!") This is the prevention strategy Francis Collins alluded to in his *New England Journal of Medicine* article, in which the hypothetical patient mustered the motivation to quit smoking when "confronted with the reality of his own genetic data."

In the past few years, the popular culture has seen a wide increase in media stories and companies putting forward the notion of using genetic information to personalize lifestyles and become a proactive player in one's own health. A company that goes by the name of My DNA Fragrance will, for instance, produce perfume from your DNA, because, as the website says, "Every person's DNA blueprint is different. So no two fragrances will smell the same." The health side of the equation is even broader. You can order (allegedly) genetically personalized sports drinks, nutritional supplements, and exercise programs. Genetics is being sold as a path to an individualized, healthy lifestyle, not only by almost all direct-to-consumer genetics testing companies, but also by many researchers working in the field. The message of Navigenics, the biggest competitor to 23andMe, exemplifies this pervasive prevention ethos, in that they claim that their "goal is to empower you with genetic insights to help motivate you to improve your health." They also claim that they can provide you with information to answer questions about what diet, exercise, and medical tests and screening will "help you most" and are "really right for you."

There are, as I see it, two problems with the idea of using genetic information for the purposes of prevention and/or personalizing your lifestyle. The first we have already touched on—the research simply hasn't uncovered genes that provide highly predictive information. The second relates to what we, as individuals, do with this information, a point I'll elaborate on momentarily.

The idea of personalizing your lifestyle with the help of genetic information looks to be almost as big a bust as gene therapy. In a 2008 article in the *New York Times,* David Goldstein, a geneticist from Duke University, stated: "There is absolutely no question that for the whole hope of personalized medicine, the news has been just about as bleak as it could be." He went on to note that "after doing comprehensive studies for common diseases, we can explain only a few percent of the genetic component of most of these traits ... It's an astounding thing that we have cracked open the human genome and can look at the entire complement of common genetic variants, and what do we find? Almost nothing. That is absolutely beyond belief."

Yes, it is beyond belief, especially when you consider the ongoing scientific push for and media profile of the idea of prevention.

Numerous other academic and policy groups have come to the same conclusion as Goldstein. A 2008 study, for instance, published in the *American Journal of Human Genetics,* concluded that "the scientific evidence for most associations between genetic variants and disease risk is insufficient to support useful applications ... it could take years, if not decades, before lifestyle and medical interventions can be responsibly and effectively tailored to individual genomic profiles." Many other studies in recent years have hinted that we may never find genetic risk information that will be sufficiently robust to be of clinical value. For example, in 2010, researchers looked into the utility of genetic risk factors as compared with the old-school methods. To their surprise, the researchers "found that a genetic risk score using multiple markers for cardiovascular disease offered no significant improvement in the prediction of the disease over traditional risk factors, such as total cholesterol and blood pressure levels."

I set up an interview with David Goldstein, the Duke University geneticist mentioned above. In person he sounded only slightly more optimistic than he had in the *Times* article. In response to my question about where the field of genetics is heading, he started with a caution: "Predicting where science will take us in the near future is almost always a bad idea."

New genome research techniques, such as those being used by Goldstein's research team, will lead to the identification of a "small number of individuals with high-risk genes." But Goldstein, much like Tom Hudson, is unsure about what we'll be able to do with that information. Goldstein told me that the public face of genetics, the idea that we can all access useful health information, has it backwards, in that it has flipped the reality of what's actually taking place.

"The idea of personalized genomics has run in the opposite direction of what the research actually says," he said. In fact, gene chip technologies like those used by 23andMe "can only provide information on low-risk mutations with little meaning. Researchers will find a small number of high-risk mutations, but we can't do anything to help these people. Even if we can make those predictions, can we do anything to help those identified at high risk? We don't have any particularly encouraging evidence."

This leads me directly to the second major problem with the notion of using genetic information either for prevention or for lifestyle adjustment. Goldstein's rhetorical question ("Can we do anything to help those identified at high risk?") is just a different way of asking how successful we can ever be at getting people to adopt healthy habits. Will providing individuals with genetic risk information help change bad habits? Will knowing I'm at a slightly increased risk of getting a particular disease be enough to get me on a treadmill or to cram some broccoli down my throat?

Colleen McBride, chief and senior investigator for the Social and Behavior Research Group at the National Human Genome Research Institute in Bethesda, Maryland, is one of a handful of scholars worldwide looking at how genetic information impacts health behaviour. Her team hopes to come up with data that will move the prevention debate beyond speculation, a point she made, in the simplest of terms, at the start of our conversation. "We need to bring evidence to these debates."

In response, I suggested that the evidence currently available tells us that genetic risk information isn't very likely to change behaviour.

"Any risk communication expert would laugh at the suggestion," McBride told me. "We have long known from other areas of research that the communication of risk is necessary but rarely sufficient. And even if people do change, they all relapse."

In other words, you take the pounds off, you put them back on. Study after study has shown that it's difficult, if not impossible, to get people to change their ways even when faced with powerful risk information. Humans are hard to motivate, and we are, by and large, slugs. Even armed with the knowledge that a behaviour is unhealthy, dangerous, or just plain stupid, we persist. The provision of genetic risk information will do little, at least on its own, to change this reality. This has been understood for years. To cite just a few examples, a study published in 2001 in the *British Medical Journal* concluded that "the current evidence suggests that providing people with DNA derived information about risks to their health does not increase motivation to change behaviour beyond that achieved with non-genetic information." And a review paper published by McBride and her team in 2010 concluded that "genetic information based on single-gene variants with low-risk probabilities has little impact—either positive or negative—on emotions, cognitions, or behaviour."

And it's worth reiterating that most of the available genetic risk information, such as the information I received from 23andMe, is weak. Would a risk increase of 0.5 percent really cause you to stop eating potato chips? McBride's overall assessment is, "Don't go to these testing companies looking for risk information that will motivate you."

Theresa Marteau, a professor of health psychology at Kings College in London, agrees with McBride's assessment. Indeed, when I talked with her she was just finishing up a systematic analysis of all available research on the behaviour change issue. The review was published in fall 2010 as part of the influential Cochrane Collaboration, an international network that produces evidence-based reviews in order to inform health-care decisions. Her paper comes to the same conclusion as McBride's. There is no evidence to suggest that communicating genetic risk information will have a major impact on behaviour, despite the claims and hopeful predictions to the contrary.

"People get powerful risk information all the time—from the bathroom scale, from the blood pressure cuff—and they do nothing," Marteau told me. We discussed her research during a break at an international workshop convened to explore the value of genetics to public health. "Waist circumference and BMI [body mass index] are probably a better predictor of type 2 diabetes than any genetic tests information." (Indeed, research has confirmed Marteau's pessimistic, off-the-cuff appraisal of the predictive power of genetics. A 2008 study found that looking at 18 genetic variants that were predictive of type 2 diabetes was only marginally more predictive than simply looking at weight, age, and sex.) "Genetic information is high-tech," she said during our meeting. "It's exciting and it seems like it ought to help. But, to be honest, providing people with risk information has done very little."

In fact, research has shown that people often don't change their behaviour even when faced with *extreme* risk information. A 2010 study found that only 37 percent of cardiac patients were engaging in recommended exercise a year after a heart attack, bypass surgery, or angioplasty. In other words, some patients wake up, see a scar on their chest in the bathroom mirror, and *still* don't change their behaviour.

There are a couple of other nails to hammer into the coffin of the prevention idea (perhaps not the best metaphor when discussing risk information). Consider this: if genetic risk information could change behaviour in a positive direction, might it not also push people the opposite way? For every individual at an increased risk for a disease, there will be someone at a decreased risk. (In fact, my 23andMe test seemed to show more "decreased risk" genes than genes that increased risk.) A person might think, quite incorrectly, "It says here I have 0.5 percent less chance of getting heart disease. Bring on the fries and gravy!" If we were to accept that genetic information could change behaviour for the better, we would be no further ahead. Some people would make positive changes while others would order an extra poutine.

Finally, there is also the possibility that some people will adopt a fatalistic attitude to a cautionary genetic indicator. They might believe, incorrectly, that a genetic predisposition to a disease means that their fate is sealed. That they can't do anything to avoid the inevitable, so they might as well indulge in life's guilty pleasures, not unlike the two fellows I met in the bar who took gym breaks to go for a smoke. Here is another wonderful example of this kind of reaction. In the late 1990s researchers found that a population in Newfoundland had a predisposition to heart disease. Did the community embrace a rigorous fitness culture? Did they turn their backs on the celebrated Newfoundlander tradition of eating when you're hungry and drinking when you're dry? On the contrary, they viewed the information as encouragement to enjoy life. As reported by the lead researcher, their initial reaction was: "This is great ... This means we are doomed so we might as well live it up. We don't need to quit smoking or change our diets."

This was perhaps not the reaction researchers expected, or that someone like Francis Collins might have hoped for.

The disappointing reality of the "genetic revolution" seems to come down to this: there have been few gene therapy successes and a similar scarcity of highly predictive genetic tests for common diseases. There is a pressing need for decades of expensive clinical research to reveal what, if any, benefit will be derived from genetic testing. There is little evidence to suggest that genetic information motivates anyone to adopt preventative health strategies. And, despite all this, we are still constantly told that a genetic revolution is underway. Even as I was writing this book, Francis Collins published an editorial in *Nature* claiming that the genetic revolution has arrived, citing as evidence the existence of predictive tests, preventative strategies, and primary-care physicians who practise genetic medicine.

Where is this revolution happening? Not in the world I live in. How, I wondered, can Collins's view so dramatically conflict with the available evidence? And how come we still see headlines that claim scientists have found genes for laziness, popularity, and getting lost?

Jim Evans is a clinical geneticist and researcher at the University of North Carolina (UNC) who has keenly followed the popular portrayal of the genetic revolution. He's chaired various US federal policy committees on genetic issues and has conducted research at UNC on topics such as direct-to-consumer testing. To Evans, there are a number of factors that contribute to the hype.

To begin with, he says, "we always gloss over the stuttering, uneven pace of science which includes dead ends and wrong turns. This doesn't negate the value of the research endeavour. On the contrary, we will be better off for doing this research. But science is an unpredictable, long, circuitous slog, and we always seem to forget this." Evans has also noticed shifts in the promises that are made. "There has been an erosion of claims in the face of reality. The claims have become less and less grandiose. It used to be gene therapy. Then it was highly predictive genetic tests. Now we have a focus on behaviour change. But, trust me, behaviour modification is going to be a bust too."

The other big factor that has led to the exaggerated claims made for genetics, according to Evans, has been the rapid pace of technological advance. The ability to sequence entire genomes—a genuinely amazing accomplishment—has created an unattainable level of expectation. "The basic science," he told me, "is moving much faster than we could have imagined. This sets up an expectation that we will reap benefits in patient care, but I'm skeptical that we'll find robust genetic predictors of the kind that will make any difference for the individual."

David Goldstein agrees with Evans. Scientists often blame the media for overselling the benefits of their research, but, says Goldstein, "scientists have a lot of responsibility for the hyped portrayal of genetics. Scientists are under pressure to get research resources to come to them instead of the next guy. The genetics community wants to make it look like we are on course to help with common diseases, even if we aren't. If we don't create that impression, the money might go to another area, like stem cells."

Geneticists, says Goldstein, should rein in their own rhetoric in order to avoid "an inevitable backlash." The popularity of his position among his colleagues was made clear to Goldstein recently when a very prominent genetic researcher, whom he declined to identify, told him to "shut up or the money will go somewhere else."

There is, to be sure, unrelenting pressure on researchers to make their work appear sexy and current, and to make it seem broadly relevant and immediately valuable. The competition for funds is fierce. Researchers and research institutions are under constant pressure to justify their work in terms of near-future economic and therapeutic benefits. In the knowledge-based economy, university research is viewed as an important part of the economic agenda. In such an environment, it is only natural that results are presented in a manner that emphasizes immediate benefits and possible, near-future therapies.

Just how much pressure are scientists under to produce results that yield financial benefits? In 2010, I collaborated with my colleague Tania Bubela to answer this question. We conducted in-depth interviews with senior technology transfer staff (the people in universities who help scientists find commercial applications for their research) across the country and found a remarkably consistent message: universities face unprecedented and unremitting pressure to produce economic benefits. This push is coming primarily from government. In fact, many of the funding agencies now have "economic growth" as one of their explicit objectives. In Canada, federal legislation created our primary funder of biomedical research, the Canadian Institutes of Health Research (CIHR). That legislation states, in no uncertain terms, that the goals of the CIHR include "economic development" and "the commercialization of health research."

The majority of the people we spoke with also stated that this pressure to make money was out of sync with the universities' true and historic mandate, which they viewed as being the production of knowledge.

In an article I wrote in 2005, I suggested that this push to produce economically valuable and immediately practical research is the foundation of a phenomenon I labelled the "cycle of hype." The cycle goes like this: researchers and research institutions feel the need to be seen as producing instantly relevant knowledge and applications for that knowledge. This results in messaging (through academic papers, enthusiastic press releases from the university and the research funders, and interviews with the media) that overemphasizes benefits and underemphasizes the risks and the limitations of research. These messages are picked up by the press and transmitted to the public. The public's appetite for news about exciting science goes up and the incentive to hype is increased accordingly.

In the short term everyone in the cycle of hype benefits from the hype. The researchers and their institutions gain notoriety. Their work is justified in terms of its health benefit or economic impact, as required by funding agencies and the government. And the media gets a good story.

Of course, the cycle does not benefit the public. Our sense of the value of genetic research and the role it can play in our lives has been twisted by the hype.

The media also plays an active role in the cycle of hype. The news industry—be it TV, newspapers, or the internet—is enormously competitive. Stories must be sold not only to the public, but to editors and producers. This being the case, the media has a tendency to represent scientific events in extreme terms. Roger Highfield, science editor for the *Daily Telegraph*, suggested that the aggressive and competitive culture within the popular press "encourages triumphalism so that every gene is a milestone on that road to a cure." This is one reason we get the crazy headlines touting the laziness or party-animal gene.

Our research team explored the hype phenomenon in several studies. For example, in 2004 we published a systematic analysis of more than 600 newspaper articles on genetic discoveries. In an effort to uncover where the exaggerations were coming from, we carefully compared the newspaper stories with the original peer-reviewed science articles that were their source. While we found many instances of exaggeration (37 percent of news stories were found to have some degree of it), a surprising number were technically accurate. In fact, our work seemed to support the notion of a cycle of hype. The media is not the sole source of exaggeration and inaccuracy, as is often suggested by the scientific community. We speculated that much of the hype comes directly from researchers and research institutions and is then transmitted, relatively faithfully, to the general public via the media. Indeed, we concluded that "the media, scientific journals and the scientific community at large may be inadvertent 'complicit collaborators' in the subtle hyping of science stories."

The fact that scientists and the media sometimes hype the potential benefits of genetic research may not come as a surprise. But, rest assured, there are more insidious and controversial forces at work.

Recent research has highlighted the degree to which a range of powerful—and not exactly health-oriented—industries have invested in and promoted genetics research, most notably the tobacco, alcohol, gambling, and even junk-food industries. Why would they want to promote a hyped version of genetics? Two related reasons. First, by emphasizing genetic causes, these industries can seek to carve out a sector of the population that should avoid their product. If, for example, they can show that a mere 10 percent of the population has the gene for an increased risk of addiction to nicotine or of lung cancer, the rest of the population could theoretically smoke with impunity. Or so the addiction industries could argue. Second, a focus on a reductionist, genetic account of addiction and disease might help to undermine broader public health initiatives, such as taxes and bans, which are aimed at reducing consumption of these unhealthy products.

Such suggestions are not just the speculations of a conspiracy theorist. A recently published paper by an interdisciplinary team of scholars from Harvard Medical School, the University of Minnesota, and the Mayo Clinic concluded that the tobacco industry "has for decades sought to use genetic information for its own profit." The authors speculate that Big Tobacco has and will continue to use genetic research "to its own ends, changing strategy from creating a 'safe' cigarette to defining a 'safe' smoker."

Other addiction industries have employed similar tactics. As noted in a 2010 paper by Australian researchers Hall, Mathews, and Morley, the alcohol industry has also "promoted the idea that alcohol-related problems only occur in a minority of genetically vulnerable drinkers." This approach benefits booze manufacturers by absolving them of responsibility for the addiction of some of their customers. Consequently, to quote the authors, "alcohol problems are better addressed by identifying and intervening with problem drinkers rather than adopting effective strategies for reducing population-level alcohol consumption."

No one is saying that such strategies have had a direct impact on the perceptions of every genetic researcher. But these vested interests have certainly, and knowingly, helped to build the momentum around the idea that genetics could be used as a way to prevent common diseases. As Helen Wallace contended in a study published in 2009, "The tobacco industry has played a significant role in shaping research agendas, in particular, by promoting the idea that individual genome screening would be of benefit to public health."

I ran into Wallace just after this study was published. She is the executive director of a not-for-profit organization called GeneWatch UK, which monitors developments in the area of genetics from the perspective of the public interest. She has a PhD in applied mathematics, has worked in both the academic and the policy-making world, and possesses a rare combination of personality traits: she is a fiery and fearless intellect but also warm, fun, and approachable.

Right from the start of the genetic revolution, Wallace was dubious. Unlike me, she never bought the hype. "It seemed incredible to me that they [genetic scientists] were claiming that adding in genetic factors would make it possible to predict which individuals would get complex diseases like heart disease and cancer," she told me in a correspondence that followed our initial meeting. Given this skepticism, she has long wanted to know: "Who's selling this story?"

A key vendor of the hype, her recent research concluded, was the tobacco industry, especially in the early days of the genetic revolution. From 1990 to 1995, 52 percent of all research projects funded by British American Tobacco (an industry group that represents over 250 brands) went to genetic research, mainly based in universities.

"The industry spent hundreds of millions, probably billons, of dollars overall," Wallace told me. "But it also infiltrated scientific institutions and even cancer charities and influenced what was funded and published by others. In the documents, tobacco-funded scientists boast of starting a race to find the genes for lung cancer."

Wallace insists that this funding and support have had an enduring impact. "Do not underestimate the role of the tobacco industry in creating a 'scientific bandwagon' that focused on identifying inherited genetic differences as the leading cause of cancer." (The idea of a "scientific bandwagon" describes the situation where a large number of researchers and research institutions commit their resources to one approach to a problem.) "The effect of this scientific bandwagon," Wallace continued, "was to shift billions of dollars into genetic research and away from research into environmental, social, and economic factors in disease."

Once the bandwagon got momentum, it kept rolling. And people are jumping on it to this day.

So, what does all this mean for the rest of us? We need to be aware that there are forces twisting the message about genetics. They include businesses (some with a test to sell) driven by the need to make a profit. They also include scientists who need to sell their area of research in order to secure funding and members of the media who are under pressure to find and then sell a good story. On top of all this, there has been a de facto selling of the entire research area by industries with a destructive agenda. Only by becoming aware of these forces can we critically appraise the role of genes in our own health. Unless you have a rare single-gene mutation that causes a specific genetic disease (such as Huntington's disease or cystic fibrosis) or one of the few genetic mutations that are highly predictive of a common disease (such as breast cancer or colorectal cancer), the research to date tells us that the role of genes is real, but hardly dominant.

The misrepresentation of the science can also create unrealistic expectations that fuel the premature implementation of new technologies. Indeed, many commentators believe this is exactly what has happened in the context of genetic testing. As Jim Evans said, the science has moved so quickly that there is an expectation of a clinically useful product. But, with a few exceptions, this simply is not the case. Add the pressure to commercialize research, and you have a recipe for the too-hasty translation of basic research.

Having said all that, I do want to emphasize that genetics remains an incredibly exciting and vitally important area of research. It has provided us with new insight into the biology of numerous diseases. It has helped to explain the evolutionary and migratory history of the human species. And it has and will continue to inform the development of new, more targeted and individually tailored therapies. But genetics is just one piece of a complex puzzle. It is part of the story, not the whole story. Most of the applicable risk factors are truly so small that the cliché "don't sweat the small stuff" really does apply. As we have seen in the last two chapters, if you are concerned about your health, it makes more sense to pay attention to your diet and level of exercise than to your genetic predispositions.

I was conflicted about my own genetic testing experience. I knew that much—okay, practically all—of the information provided by 23andMe was useless from the perspective of health-care decisions. Still, I found it intriguing. I loved the fact that I had the *longevity, good memory,* and *slightly-above-average height* genes (even though the research supporting the relevance of these genes is still thin). The ancestry aspect of the testing, which 23andMe also offers, was compelling. It turns out that I am, as Joanna Mountain put it, "Irish to the core." I don't have a single marker from any location on planet Earth save the Emerald Isle, though it remains beyond me how this explains anything except my love of Guinness. Other than that, visiting 23andMe provided me with no information I was able to envision using to improve my health in any meaningful way.

But ... there was still that one shocker, the one piece of genetic information that gave me so much anxiety and made me swear out loud, causing my children to stop in their tracks. Ironically, this bit of information had nothing to do with my health. Here I must caution that I am about to be somewhat immodest.

I was about five years old when I realized I was fast, the fastest kid in every grade, which said a lot because we moved often when I was young. I held various school and county records, and at the age of 12 I joined a track club and trained with Olympians, nakedly aspiring to be one. My passion continued throughout high school and university. Although I won or did well in some big races, it eventually became clear, painfully clear, that I wasn't Olympic material.

It didn't matter, because the sport gave me much: my first formal date; my first overnight trip away from home; wild post-race parties that helped to establish close friendships; my wife. I met Joanne, a world-class runner with all the right genes, at our track club. After university I shifted to sprinting with a bike on a velodrome, a sport I still pursue competitively. I could continue, but you get the idea. Sprinting has been central to my personal history. I am a sprinter, through and through.

Or am I?

On that busy night at home when I first scanned the results from 23andMe, one item jumped out. One 23andMe analysis is of the genes that code for "muscle type." The test result said the following, in plain language: "unlikely sprinter." My genes did not possess the code for quick-twitch muscles. Zero, in fact. The text following the "unlikely sprinter" result also suggested I pursue endurance events.

Ouch. Damn Irish genes.

I sulked for days. I frequently found myself looking at my legs and cursing their unsprinterly muscles. In its long history, Ireland has not produced many world-class sprinters. It seemed I was now part of that unstoried tradition.

But I wondered, *Would the right quick-twitch genes have propelled me to Olympic glory?*

Probably, or more likely, almost certainly not. Some other physical failing would likely have limited my success. Short shins. Fat feet. Hairy, drag-inducing legs. Of course, if I had been born in the era of the genetic revolution, my parents could have had me tested for sporting propensities. Many companies now exist that will test your child, for as little as $150, to find out if he or she has the genes for speed or endurance. Atlas Sports Genetics, for example, tests children as young as one year old, allowing their parents to make an informed decision about which sport to frantically push, I mean encouragingly place, their children in.

Had I taken such a test perhaps I would have become a famous endurance cyclist. Or perhaps I'd have been just good enough to be distracted from university and would now be working the late shift at a convenience store. Maybe I would have detested the endurance sports I am apparently genetically suited to pursue and would have done nothing but sit on my posterior. Most likely, due to the constellation of complications that make us human, I would not have been any better at climbing a mountain on a bike than dashing 100 metres down a track.

A few weeks after my visit to 23andMe, I ran into Jim Evans at a conference. With great sorrow, I told him of my non-sprinting-gene affliction.

"The fact that you don't have a quick-twitch genetic predisposition but still enjoyed and excelled in sprinting just shows how complex things like sports are," he said. "But it's more profound than that. The greatest thing about having evolved is that we're free from being slaves to our biological destiny. We can violate the imperatives of biology. We can pursue activities regardless of various simplistic deterministic predispositions." Evans paused a moment before continuing, and then smiled. "That's how we find joy."

I'll raise a Guinness to that. What else can I do?

4

REMEDIES
BIG PHARMA AND THE COLON CLEANSERS

Puke. Upchuck. Vomit. Throw up. Hurl. The liquid scream. Painting the pavement. Emesis. And the more delicate, though descriptively incomplete, *unsettled stomach.* There are countless terms to describe the act, one of the least enjoyable in the human repertoire of bodily functions and one of the most unambiguous signs of illness. And it is an act I have performed all over the globe. I have thrown up in a variety of countries, climates, and cultural surroundings. Why? Because I have a strong propensity toward motion sickness.

Motion sickness has compelled me to leave my gastric mark—on the street, in an alley, or, after a desperate sprint, in my hotel room—in Doha, Rome, Sydney, Tasmania, Tuktoyaktuk, Munich, Mexico City, London, Oxford, Paris, Tokyo, Dhaka, the Great Barrier Reef, and Los Angeles. The winding road from Vancouver to Whistler has cut me down on several occasions. Ditto the trip in from the Montreal airport, which I attribute to the spirited approach to driving embraced by the local taxi drivers.

At times, the act has been quite dramatic. On a visit to Taipei, two enthusiastic and accommodating graduate students offered to take me to see the stunning National Palace Museum. Since neither of my hosts could speak English, I did not realize that the trip to the museum would start with a two-hour drive through the (also stunning) hills surrounding the city. Despite my attempts to communicate my predicament, the driver forged on. I sat in silent, clammy agony until the car stopped in front of the museum. I attempted a smile toward my driver, said "toilet," and bolted from the car. I jogged toward the mountain of stairs at the museum's entrance. It might as well have been Everest. I wasn't going to make it. I searched for a bathroom. No luck. I spun around and saw a trashcan. It would have to do. I believe I was as discreet as a person can be puking in front of one of the world's great collections of historic and cultural artifacts.

On another occasion I was being transported from Ottawa to Meech Lake in a black limousine-like van. It was one of my first rather official national policy events and one of the rare occasions in my career when a suit and tie were expected. As we drove along the beautiful roads that cut through Quebec's Gatineau Park, I did my best to make semi-intelligent banter with the politicians, senior bureaucrats, and researchers in the undulating vehicle. But my gills were getting greener with each passing mile. Moments before we pulled up to the conference centre, I hopped out of my seat and pressed up against the door of the van, a move that generated admonishments from the driver. Someone yelled, "Open the door, he's going to be sick." I like to think it was the federal industry minister, but I can't be sure. The door was flung open and I heaved behind a small shrub just a few metres from the van.

I've tried to keep this problem fairly secret ("vomits easily" is not the trait I most want to be known for). But, as highlighted above, it is pretty difficult to hide, and word gets around. Colleagues, friends, and family members have recommended a variety of cures. My sister-in-law Akiko is from Tokyo. Her surefire remedy, passed down from her mother, involves the placement of a pickled plum, an umeboshi, in the belly button. Another remedy, this one from Ireland, was to tie a cluster of mint to my wrist. A law professor colleague offered me a homeopathic pill as she watched me sit pasty-faced during a taxi ride from the Sydney airport. "Works right away," she assured me. Other remedies, such as ginger, acupressure wristbands, and, of course, a range of pharmaceuticals, have also been suggested.

To date, I have tried none of them.

Given the severity of my affliction, and the frequency with which I travel, you might think this absurd. Why have I avoided taking some kind of remedy when both my dignity and my stomach have been subjected to such regular assault? The plain answer is that I am deeply skeptical of all unproven, alternative approaches to health, and equally skeptical of pharmaceuticals and their inevitable side effects. A plum in the belly button may be quaint and harmless, but it is also absurd and probably sticky. I also hate being drowsy, which is a common side effect of the medications most often recommended. Not to mention that the pharmaceutical industry's actions over the past couple of decades have not, to put it mildly, generated confidence in the efficacy and safety of their products.

My reaction to the remedies for motion sickness seems to map the perceptions and challenges associated with almost all remedies. (For my purposes, a remedy is something that is meant to cure, or lessen the effect of, a disease or bodily disorder.) Remedies either have an active agent or invasive process that can cause both benefits and harm (pharmaceuticals would be the obvious example here) or involve some nebulous or unproven force that has no scientifically verified foundation and, consequently, little hope of doing anything (such as a strategically placed pickled plum). Both types of remedies have attractive qualities. Pharmaceuticals are, at least in theory, backed by the weight of scientifically derived evidence, are standardized, and are often prescribed and administered by highly trained, certified, and regulated health-care professionals. Alternative medicines, as they are often called, come with the promise of a personalized and holistic approach to care, relief with no side effects, and, for some alternative approaches, a spiritual component.

To some readers, my categorization may seem a bit rough. Some of the remedies typically labelled alternative, such as herbal remedies and acupuncture, clearly have the potential to do *something* to the body. A needle stuck into the skin is not a biologically benign act. A herb is a bundle of chemicals, not unlike any pill purchased in a pharmacy. And, as we will see, many of the remedies offered by the conventional medical establishment are also of questionable value, and the evidence to support their use is often clouded by controversy.

But, despite the limitations of my classification, I feel it captures the spirit, if not the technical reality, of the dichotomies we are consistently presented with in this sphere. Big business versus local practitioner. Cool empiricism versus an insight-informed personal touch. Institutional science versus traditional knowledge. Evidence-based decision making versus intuition and spirituality. And, one of the most pervasive and nonsensical oppositions, unnatural versus natural. Though all are factually false (alternative medicine, for example, *is* a big business) or misleading (what exactly is "natural" and why is that so healthy?—arsenic is natural after all), these binary classifications have framed the debates over health remedies.

In the preceding chapters we have explored three major factors that influence our health: fitness, diet, and genetics. But what about when we are not healthy? We all get sick or have some health problem that we need (or think we need) to mend. When this happens, we turn to a remedy. For the next part of my journey I will explore an array of remedies. This is yet another mind-bogglingly vast subject. If you include the relevant research, the manufacture and sale of medicine and associated products, and the education of health professionals, then the provision of remedies amounts to one of the largest sectors of our economy. In wealthy nations, remedies are an obsession. Our world is filled with advertisements for remedies for everything from general malaise to sexual dysfunction to serious, though likely incurable, diseases. No matter what the problem, we want to get better. We want to be fixed. We want a remedy.

For this chapter, I am going to keep it simple. Using one of my (many) frailties, motion sickness, as the primary focus of my investigation, I will show that, in the context of remedies, the forces that mislead are remarkably varied. These forces are financial and philosophical, understated and overt. And they make it very difficult to believe anything.

The clinic was in a renovated old house in a pleasant part of the city core. Most of the nearby buildings were similarly revamped residential homes converted to office use, and all exuded a cozy and hip but still professional vibe. If I didn't count the time my chiropractor friend cracked my back on a park bench (which resulted in tingly pain down the length of my right arm), this would be my first visit to an alternative health practitioner.

The clinic itself had a pleasant feel, the decor a perfect mix of "doctor's office" and "welcoming family home." This blend seemed ideal, in that it paralleled the ethos of the health professional who worked there. Naturopaths are becoming an increasingly popular health-care option for Canadians and, to a lesser degree, Americans. After my visit, I understood why. The naturopath who saw me was fashionably dressed, cheery, and on time, a rare phenomenon in the health-care universe. We made small talk as she led me to the examination room, though it looked more like a big living room with a desk than the cubicle found in most doctors' offices. No white walls, industrial metal chairs, or graphic educational posters. I told her about my primary concern, my motion sickness. She took careful notes as we sat and calmly discussed my life and lifestyle for more than 45 minutes. She asked about my sleeping patterns, if I had a good relationship and sex life, how much exercise I got, if my digestive tract was working well, the nature of the stress in my life, and what I ate and drank. (Given that I was still in the midst of my diet program, she was impressed. "You get a gold star for nutrition," she said, smiling.)

This encounter was the most pleasant clinical experience I have ever had. No exaggeration. It was actually enjoyable. How often can you say that about a visit to the doctor? The practitioner appeared both skilled and knowledgeable. She gave the impression that she was listening attentively to my problems and thinking carefully about an appropriate solution. I had no complaints at all about her bearing and attitude.

But what about her solutions? She recommended deep breathing exercises, several homeopathic potions, and a visit to an acupuncturist.

Before I walked in the door I had reminded myself to keep an open mind. But I felt my skepticism hovering nearby as I scanned the magazines with natural health themes piled on the waiting room credenza. Yet within minutes of sitting down with this caring practitioner my uncertainty dissipated and I found myself absorbed in a give-and-take dialogue about my health and lifestyle. Now, to be fair, my satisfaction was likely fuelled by the fact that this was a 45-minute discussion about me. Who doesn't like talking about themselves?

Feeling oddly invigorated, I left the naturopath's office with a referral to an acupuncturist and a bag containing a homeopathic solution, some vitamins, and fish oil. I was also out of pocket by about $250.

A few days later I showed up at the acupuncturist's office. While I had to wait 20 minutes to see him, the quality of the clinical experience was no less agreeable than my trip to the naturopath, chiefly because, once again, he seemed fully engaged in my health needs. I explained my primary concern, motion sickness, and he went straight to work.

The diagnostic process administered by this professional, who was articulate, witty, and poised, did not instill confidence in the potential effectiveness of the modality. He briefly looked at my tongue and distractedly took my pulse. I must have given him some sort of look because he assured me that that was what his training, which was in traditional Chinese medicine, required him to do. Armed with this limited data, he divined that my "meridians"—the channels through which my life energy flows—were poorly aligned and that the energy flow to my stomach was particularly out of whack, hence the motion sickness.

Once we got to the needles, which I was dreading, he radiated confidence. This, I thought, was a man who knew exactly what he was doing. Moving and talking with agility and purpose, he explained the ancient Chinese meridian system as he had me lie down on a big exam table. He then proceeded to insert needles into my body. He started with six in my face and ended with many more in my feet. My whole body was covered. The standard pincushion metaphor is not out of place here. As he was inserting needles into my flesh we talked about topics ranging from provincial politics to the lunacy of various 9/11 conspiracy theories. Whenever I spoke I made the long handles of the face needles wag. It didn't hurt, but it was certainly an odd sensation.

Moments after he extracted the last needle he took my pulse again, for perhaps two seconds, and informed me that the energy flow to my stomach had already improved. Meridians aligned and $125 poorer, I was good to go.

Although a touch New Agey, these two sessions had been professional and even enjoyable, but their effectiveness had not yet been gauged. Had the remedies provided by these alternative practitioners cured or at least alleviated my motion sickness? As it had with my diet, the Alaskan cruise would provide a handy test. Nothing makes me sicker than a boat ride, and my visits to the naturopath and acupuncturist occurred just days before the floating buffet was scheduled to depart. I vowed to follow all the advice provided by my alternative practitioners: I would take the homeopathic remedies in the prescribed manner and, if necessary, apply the on-the-spot acupressure moves that could be self-administered to alleviate nausea (the acupuncturist had demonstrated several, including finger pressure on the wrist and shin). My seagoing plan was to avoid any pharmaceutical, no matter how queasy I felt, unless I actually vomited. If and when I hit the vomit barrier, only then would I resort to a conventional pill.

I realize this was hardly a scientifically valid experiment. It was methodologically flawed on so many levels that it was unlikely to produce even compelling anecdotal results. There was one research participant, me. I was mixing remedies, including a homeopathic solution, deep breathing exercises, and pre-aligned meridians. And my cruise-ship experiment was not blinded—that is, I knew what remedies I was using. Most significantly, however, I entered the experiment with a predisposition to be skeptical of both alternative and conventional remedies. Under these conditions, the chance of my experiencing a placebo effect was minimal. Nevertheless, if these treatments had a clear benefit, I sensed I would be able to detect *something*. If, despite the imperfections of my hang-in-until-I-vomit experimental design, there were a powerful therapeutic effect, it should, at least, ease the queasiness.

Over one-third of North Americans regularly use some type of complementary and alternative medicines (CAM), and it's likely that over 50 percent have tried a remedy—herbal medicines, chiropractic treatments, therapeutic massage—in the past year. Its use is on the rise among children, via their parents, and in virtually every developed nation. Numerous studies have found that a variety of patient populations, such as those with cancer, use even more alternative remedies than the general public.

In short, alternative medicine, despite a name that implies a place on the periphery of health-care choices, is a mainstream phenomenon. It is not alternative; it is the norm. It is a multi-billion-dollar industry (close to $40 billion for alternative medicine products in the United States alone) largely supported, not by individuals from fringe cultural communities with unorthodox world views, but by wealthy, white, educated women. Studies have consistently shown that middle-aged women from the upper socio-economic strata are the demographic mostly likely to use alternative medicine. When you think *alternative medicine,* think SUVs and suburbia, not ancients and mystics.

In North America, the most popular forms of CAM include three of the practices I tried: acupuncture, homeopathy, and naturopathy. Chiropractic, a practice founded on a belief that the subluxation (misalignment) of the spine can cause a range of ailments, is also among the most popular. But given my one arm-numbing experience with a chiropractor, I could not bring myself to expose my joints to manipulation, even though I'm sure I could have found a chiropractor willing to take a stab at administering a cure for my motion sickness.

Naturopathy has a long history. It emerged out of the "nature cures" movement that thrived throughout Europe in the 1800s. A dominant figure in this movement was a German, Sebastian Kneipp, who believed that a kind of water therapy—basically cold baths, hot baths, and "gushes" or blasts of water from a watering can or hose—had cured him of an ailment he had acquired by studying too hard for the priesthood. The water cure brought back his vitality. It worked, Kneipp supposed, because the water removed the bad substances, the "morbid matter," from the body. Kneippism, an approach to health built around the water-cure method, became fashionable throughout Europe and led to the establishment of a number of spas. Kneipp's primary spa, Bad Worishofen, still exists and remains a popular destination for individuals seeking his cure.

Benedict Lust was also cured of what he believed was a grave illness by Kneipp's water-based remedy or, as it is now usually called, "hydrotherapy." He became a devotee of Kneippism and other natural remedies, moved to the United States, and, in 1901, opened the first school of "naturopathy," a term he selected to describe his expanded view of Kneippism, which, in addition to hydrotherapy, included other non-drug-based and "nature"-oriented approaches, such as chiropractic, herbal remedies, and homeopathy. Thus naturopathy was born. And though its popularity has at times waned, it has been around ever since. In the mid-twentieth century, when conventional medical researchers were making big, scientifically informed breakthroughs (antibiotics, vaccinations, open heart surgery, and so on), naturopathy faded. But the New Age vibe of the 1960s, coupled with a widespread decline in respect for authority and the 1990s postmodern, all-knowledge-is-relative attitude, helped to make naturopathy popular again. There are currently more than 4000 licensed naturopaths in North America, which is almost a 100 percent increase from 2001.

It is important to emphasize that naturopathy is not a particular treatment modality, such as surgery. It is not a constellation of treatments that focus on a particular ailment or a specific part of the body, like physiotherapy or optometry. Nor is it a systematic and testable approach to understanding human biology and the disease process, like, well, science. Naturopathy is, rather, a world view. It is a philosophical approach to health and, as a result, can include the use of any remedy that fits within the belief system. While naturopathy has evolved since the days of Lust, the basic tenets remain the same. At its core, and as implied by the name, naturopathy is largely focused on the healing powers of nature (*vis medicatrix naturae*) and a belief in the inherent self-healing powers of the human body. Lust and his followers took these beliefs to the extreme, eschewing anything that could be viewed as an unnatural contaminant, such as pharmaceuticals and vaccines. (A recent study of Canadian naturopath students found that distrust of vaccines remains a feature of the practice.) Lust believed, for example, that "drugs have no place in the human body" and that their administration could be viewed as both "ignorant" and "criminal."

Naturopathy also has mystical and, in the early days, strongly religious underpinnings. The movement that gave rise to its birth—the European "nature cures" trend—was largely founded on a belief in an amorphous vital force present in all living things. Early naturopathic leaders, including Lust, gave naturopathy an overtly Christian stamp, suggesting that its methods were tied to the natural state Adam and Eve enjoyed before original sin. At one point, Lust suggested that naturopathy was part of a spiritual upheaval akin to the birth of Christ and the Reformation.

This unwavering faith in the powers of the vital life force found in nature is a nice enough idea, but it's one without any scientific foundation. Its practitioners have come up with a variety of bizarre treatments over the years. These have included wearing "Porous Health Underwear" (basically, briefs, bras, and bloomers made from natural materials that were permeable enough to allow the invigorating weather—the colder the better—to reach the private parts) and eating sand as a cure for constipation. (You might think sand would be more clogging than cleansing, but I confess to having no first-hand experience in the matter.)

It's all too easy, I admit, to find absurd remedies from the history of any health field, including medicine. (Bloodletting, for example, was practised for about 2000 years.) And conventional physicians continue to provide remedies of questionable value that exist largely as a result of a corporate drive for profit. I highlight these early naturopathic remedies and the philosophical foundations of the practice to make a point. Naturopathy is not, at its core, either scientifically informed or evidence-based. Naturopaths offer some sensible advice—eat well, get lots of sleep, and exercise—but this does not make naturopathy evidence-based. These practices are recommended because they align with naturopathic philosophy, not because they have satisfied some rigorous scientific, naturopathy-led inquiry into their value.

The naturopath philosophy—in particular, the belief in a vital life force and the healing power of nature—distorts both the use and presentation of the evidence about naturopathic remedies. We have seen that food companies represent the evidence about exercise in a manner that promotes the consumption of empty calories. Genetic researchers inadvertently (or, perhaps, consciously) twist the truth about the limits of genetics in order to promote their area of research. Companies that promote fad diets underplay the truth about their limitations. And, as we will see, pharmaceutical companies exert a powerful force on what the evidence says about their products. But the twisting effect of blind faith in a particular world view, such as the healing powers of nature, is no less severe. It too leads to spin.

James Whorton, professor emeritus at the University of Washington School of Medicine, has spent his career studying the history of alternative medicine. His book *Nature Cures: The History of Alternative Medicine in America* has been an invaluable resource to many, including myself.

"There is understandable suspicion [held by those in the scientific community] that naturopaths are under a philosophical constraint," Whorton told me. "Are naturopaths so wedded to a 'nature cures' approach that there is an uncritical acceptance of any remedy that accords with this world view? This seems likely. This philosophy prejudices and twists the evidence. If the scientific research tells us that a remedy does not work, will the philosophy allow the naturopath to see the reality of the situation?"

If one looks at the types of remedies currently provided by naturopaths, the answer to this question must be a resounding no.

Naturopathic medicine has, thankfully, progressed from the days when patients were told to eat sand and wear porous undies. And naturopathic associations throughout the world claim to have an increasing, though qualified, interest in scientific research. At the same time the practice remains, at its core, based on a non-scientific philosophy. The guiding principles for naturopathy, which can be found on the website for the Canadian Association of Naturopathic Doctors, include references to concepts that reflect the degree to which the practice remains tied to Lust's early vision, including mention of an "inherent healing ability of your body, mind and spirit."

The practitioners of naturopathy, as with many alternative approaches, want it both ways. They want the legitimacy, mainstream acceptance, and prestige that come from a perceived science-based approach to health (hence the frequent use of scientific-sounding terms and trappings), but they must also continue to embrace a defining philosophical framework. As noted by Whorton, it is very difficult to be both evidence-based *and* tied to a mystical philosophy. When the two come in conflict, which perspective will win? The answer, inevitably, is that the philosophy is paramount: if this were not the case, then the very reason for the existence of the alternative practice would be eroded. Indeed, the website for the Canadian Association of Naturopathic Doctors implicitly recognizes this reality. It states that the profession "recognizes the value of research" (this needs to be explicitly stated?) and "seeks to make appropriate use of science to further the understanding and advancement of naturopathic medicine." And what if science does not advance naturopathic medicine? It gets ignored.

I have experience of my own in dealing with this kind of contradictory thinking in relation to naturopathy. In 2009 there was a public debate about whether to grant naturopaths, who must be licensed in British Columbia, further powers to diagnose and treat ailments such as allergies. I wrote a commentary with a few colleagues for the *Vancouver Sun*. We cautioned against expanding the legal scope of practice for naturopaths and pushed for an evidence-based approach to health-care decisions. (In the end, the naturopaths succeeded in changing the relevant provincial guidelines. So much for evidence-based health-care policy!)

In a response to our article, also published in the *Sun*, the head of the British Columbia Naturopathic Association called our article "misinformation of the worst kind." He wrote: "The science behind naturopathic medicine is substantiated by voluminous research conducted by independent, third-party medical experts. In fact, the science behind naturopathic and standard medicine is not different; it is the philosophy behind the application of that science that differentiates naturopathic doctors (NDs) and medical doctors."

The claim that naturopathic medicine is substantiated by the kind of research he describes is false. Regarding the rest of his statement, it's hard to argue with the proposition that it is the philosophy behind naturopathy that distinguishes it from more science-based approaches. This is my point.

I had made it three days into the cruise without throwing up. I regularly felt queasy, but despite some undulating nights, I was faring better than anticipated. Sure, there was the constant low-grade headache, but the vomit threshold had not been crossed. I also suspected that it may not even have been the ocean swells that were causing my headache. Maroon carpeting, Vegas-style stairways, glass, chrome, and Muzak at every turn may have been more significant factors. Is there a homeopathic remedy for exposure to poor taste?

Given that my nausea had been held in check, I was forced to ask myself: was my semi-perky state a result of my scrupulous adherence to the recommendations provided by my naturopath and acupuncturist? Or could there have been other reasons? I had faithfully followed the deep-breathing procedures, including something called alternate nasal breathing—a practice meant to purify the blood and aerate the lungs, or so claimed the information sheet provided by my naturopath. And whenever my stomach threatened a violent upheaval, I used the prescribed acupressure manoeuvres. I had also consumed liberal amounts of two homeopathic remedies. One was made from passion flower and was intended to quell the anxiety associated with my fear of making a heaving spectacle of myself. The other, cocculine, was designed specifically for motion sickness. The instructions on the label recommended five doses three times a day, but my naturopath told me I could take as much as I needed as often as I liked. *No need to worry about an overdose or side effects,* she had told me in her most reassuring voice.

Homeopathy, which was created more than 200 years ago by Samuel Hahnemann, is one of the pillars of naturopathic medicine. It has been part of the naturopath's arsenal from the beginning. It uses natural substances, such as the passion flower in my anxiety potion. It has few side effects (more on this below). And it allegedly works by stimulating the body's natural healing process. Thus, it fits perfectly with naturopathy's overarching principles.

Homeopathic remedies are based on the theory that "like cures like"—the similia principle. It works like this: take a natural substance, such as a plant or mineral, that causes symptoms similar to whatever ailment you are trying to address. You then put that substance in water and dilute it to the point of non-existence. Seriously. Non-existence. Homeopathic remedies are so diluted that the therapeutic element, such as my passion flower, is no longer present in any real sense. You would need to drink absurd quantities of it merely to absorb one molecule of the allegedly active ingredient. No matter, because proponents of homeopathy believe that homeopathic solutions become *more* powerful the more they are diluted. The process of "ultra-dilution," as it is called, somehow allows the water to hold the memory of the therapeutic substance.

So, does homeopathy work? Is the use of homeopathic remedies "substantiated by voluminous research," as suggested by the head of the British Columbia Naturopathic Association?

This is a question I put to Professor Edzard Ernst. He is a

bona fide celebrity in the field of alternative medicine. A transplanted German, he holds the chair of complementary medicine at the University of Exeter, the first position of its kind in the United Kingdom. When he accepted the post he had no intention of discrediting the field. On the contrary, he viewed it as an opportunity to gain knowledge about the true benefits associated with various alternative remedies. The more we learn, he went in believing, the better for both patients and practitioners. Further evidence of his initially favourable disposition toward alternative medicine is the fact that his position at the university was made possible by a financial donation from Maurice Laing. Laing's wife had been helped by alternative medicine and Laing thought it was time to study the field at a proper university. But almost from the get-go, Professor Ernst found that the evidence—that is, the scientifically robust data—did not cooperate. Indeed, since then, rigorous studies around the globe have rarely been supportive of alternative remedies. Ernst's career panned out much differently than he had originally envisioned. He is now viewed by those within the alternative medicine community as something akin to the Antichrist.

Professor Ernst's desk was covered with the debris of a busy scholarly career: papers, journals, books. He was hunched over what appeared to be a draft manuscript. I held out my hand, and he grumbled at me before looking up, and then said, "Remind me why you are here?" I felt like Harry Potter about to be chewed out by Dumbledore for spilling a potion. Clearly Professor Ernst was not as excited to see me as I was to see him.

I told him of my search for the truth about alternative remedies, and his mood changed. Suddenly he seemed keen to discuss both the details of his research and the personal turmoil his work has caused.

"I get so much hate mail," he said. "I have acquired a thick skin, but the personal attacks that come through my peers ..." His voice trailed off momentarily. "All I'm doing is trying to apply the principles of science to the field. You would think that the proponents would love me. The fact that they don't is telling."

Ernst's perspective on the philosophical foundations of alternative remedies accords with that of Whorton.

"One thing is clear," he said, "alternative medicine is not a field governed by rationality. It is not based on rationality. It is a religion. And because it is a religion, I've come to realize the evidence doesn't matter." He recognized, he said, that "strong financial pressures govern the science in the area of pharmaceuticals." But he thinks the distorting forces are worse in the world of alternative medicine because, as he says, "a quasi-religious perspective governs, and that is a more powerful force than even money."

But what about homeopathy? I asked him. Does it work? Could it work? Are the advocates of alternative medicine twisting the truth?

No, no, and yes.

"There have been thousands of research studies examining homeopathy. There is no evidence it works." Ernst paused and then said with finality, "It does not work."

Ernst's conclusions about homeopathy are not unique; many have long believed that homeopathy is of questionable worth. Within the scientific community, "it does not work" is the accepted view. Very few studies—of the many noted by Ernst—have shown that homeopathy is better than a placebo treatment. The few that have shown some treatment effect have, in general, also betrayed serious methodological limitations. In other words, the studies were poorly designed, making the results less than compelling.

The reality about homeopathy was recently crystallized in a report issued by the British Parliament's Science and Technology Committee. The UK's National Health Service has long covered certain homeopathic treatments. In early 2010 the government decided to look into the value of this practice. This led to a bitter public debate between those who support homeopathy, a cohort of enthusiasts who usually fall back on the give-patients-the-choice argument, and those from the scientific community, who called homeopathy an "expensive placebo," "nonsense on stilts," and "witchcraft." (It was noted by one commentator, however, that this last characterization was insulting to witches.)

The committee concluded that homeopathy does not work, at least no better than a placebo, and that the "principle of like-cures-like is theoretically weak" and "fails to provide a credible physiological mode of action for homeopathic products." In addition, the committee found that the idea that "ultra-dilutions can maintain an imprint of substances previously dissolved in them to be scientifically implausible." The committee went so far as to suggest that the idea of ultra-dilutions is so preposterous that it would be inappropriate to waste money researching the topic. The report has already led to the pulling of funding. In October 2010, for example, one regional department of the National Health Service ended its long support of homeopathy, concluding there was no evidence it worked. More regions can be expected to pull funding.

The Science and Technology Committee report also included a damning observation about the misleading statements of alternative medicine practitioners. The committee heard testimony from numerous experts, including those who practise homeopathy. These individuals, the committee stated, ignored the picture painted by the available evidence. Specifically, the committee concluded with these words: "We regret that advocates of homeopathy, including in their submissions to our inquiry, choose to rely on, and promulgate, selective approaches to the treatment of the evidence base as this risks confusing or misleading the public, the media and policymakers."

You got it. They were twisting the facts.

Many people swear by homeopathy. It is a large and profitable industry (it has been estimated, for example, that Americans spend more than $3 billion on homeopathic remedies), one that includes specialty magazines, products for pets, and remedies for every disorder you can think of—from acne, addiction, and back pain to life-threatening conditions such as cancer. Comments along the lines of "amazingly powerful tool for empowering someone to heal themselves of cancer" are not uncommon on homeopathy websites. And the internet is littered with patient testimonials extolling the natural healing powers of homeopathic solutions (often accompanied by a tirade against the evils of conventional medicine and commercial pharmaceuticals). These individuals believe that homeopathic remedies have helped them to treat or cure a variety of ailments and have done so with no nasty side effects.

The former perception can only be due to wishful thinking and the placebo effect. The latter is due to the fact that homeopathic solutions are, basically, water. There is no active agent, so there are no side effects. Unless you are allergic to water, you have nothing to fear from any homeopathic solution, no matter how much you consume. This point was dramatically illustrated in January 2010, when hundreds of people in Britain organized a protest against government funding of homeopathy. Demonstrators gathered outside drug stores and attempted to overdose by consuming ridiculous quantities of homeopathic solutions and pills. For example, one protester ate an entire carton of homeopathic sleeping pills. Did he lapse into a coma? No. Did his binge produce a sleepy yawn? No. And the UK protesters did not *believe* in the power of homeopathy, so there was also no placebo effect.

It's worth noting that the placebo effect is, without doubt, a powerful phenomenon. It generates a real biological response. One recent study from Germany used MRI technology to map the impact of a placebo painkiller on the neural activity of participants. Heat was applied to their arms and then a placebo painkiller cream was smoothed onto the skin. When the participants thought the placebo was a real painkiller, they felt less pain and a pain-suppression pattern was observed in the nerves. In other words, the placebo had a physiological impact not unlike a real, pharmaceutical painkiller.

So, placebos can work, in a sense. Researchers regularly speculate that this is how many alternative therapies give the impression that they work. But is it appropriate to base an entire field of health care on the placebo effect? Do you rely for effect on a misperception of reality? Moreover, is it ethically acceptable to lie to patients about what the evidence says about a given therapy in order to induce a beneficial placebo effect? (This has been called the "lying dilemma.") Professor Ernst does not think so. "I believe all health-care providers are governed by a medical ethics. Not just doctors. You can't lie to patients," he said.

The placebo issue is relevant to the "patient choice" argument so often put forward by those who advocate the funding and legitimization of alternative medicine. I am all for patient choice. But that choice should be informed. Indeed, consent law demands that patients be told all the material information needed to make an informed choice. And that material information includes what is known about the efficacy of a treatment, obviously. Whether the practitioner involved is a medical doctor or a registered naturopath, an offer to a patient to take a homeopathic remedy should be worded something like this: "All available evidence tells us that homeopathic remedies do not work. Would you like to try one?" This kind of disclosure seems particularly appropriate if the field claims to be evidence-based and informed by science, a claim made by those who practise modern naturopathic medicine.

Many alternative practitioners may not even have the necessary knowledge base to get informed consent. A study published in 2009, for example, found that alternative practitioners are not paying attention to the research on the remedies they use. It showed they were significantly less likely to be aware of research on alternative medicine than conventional physicians. In addition (no surprise here), they expressed much less regard for research results. But, as the saying goes, ignorance is no excuse. Efficacy is clearly "material information" as defined by existing consent law. Alternative practitioners should be aware of the relevant research and should provide patients with the truth.

If patients *were* told that a particular therapy was no better than a placebo, would they still want it? Perhaps. After all, people go to alternative practitioners for a wide variety of reasons. Many of them may not care what the scientific evidence says. I recently got into an intense driveway argument with my neighbour about the effectiveness of homeopathy (yes, I am undoubtedly my block's most annoying know-it-all). During our lively debate I rolled out all the evidence about homeopathy. But, for her, the data is completely irrelevant. She does not trust science. Homeopathy works for her because she has faith (her word). And it does not work for me because, apparently, I have no faith.

As Edzard Ernst told me in Exeter, many alternative therapies are more like a religion than a science. My neighbour, for one, would be perfectly fine with this distinction. This kind of tolerance for "magical" thinking seems to be a common attribute among users of alternative remedies. A 2008 study from Toronto, where subjects included patients with life-threatening cancer, found that a tendency toward superstition was a strong predictor of alternative medicine use. For people possessed of these beliefs, evidence may simply not hold much weight.

As mentioned, there are many other, non-evidence-based reasons people go to alternative practitioners. The personal attention, I can attest, is certainly one of these reasons, a point James Whorton emphasized during our discussion. "Historically, one of the great appeals of naturopathy is that the prescribed remedies, such as homeopathy, require taking time with [the] patient. It has never been an 'assembly line' practice. There is individual attention. This has always been part of the attraction to all complementary and alternative medicine. Also, there is an emphasis on the person, as opposed to just treating the symptoms of disease. Holistic medicine, as it has been called."

Whorton's observations fit in with my own. I almost fondly recall the 45 minutes the naturopath spent with me. Rarely, in the hubbub of modern life, does someone sit and listen to me talk about myself for so long. (I think my wife used to good-naturedly suffer through these kinds of one-way discussions, but now our kids top the list of the topics-for-conversation-when-we-have-a-moment.) My naturopath used a global, or holistic, approach. Some studies have shown that this is, in fact, the number-one reason people use alternative medicine. A Canadian study published in 2008 found that the primary motivation for heading to an alternative practitioner was the way they treated "the whole person" (78.3 percent of the survey participants picked this as the top reason). The desire to take an active role in one's health was a close second.

Whorton agreed with this latter finding. Alternative medicine has always been tied to the idea of "taking control of your life," he told me. "Since the 1960s, there has been a growing mistrust of medicine. This is associated with the idea that we are losing control of our health. The public responds to the idea of individual empowerment found in alternative remedies."

Another motivation, closely related to the notion of empowerment, has to do with the affiliated philosophies. Many people go to naturopaths because it fits with their world view. For example, they may want to align themselves with the "nature cures" ethos. Going to an alternative practitioner becomes a form of self-expression. A study published in the *Journal of the American Medical Association* found that people go to alternative providers because "they find these health-care alternatives to be more congruent with their own values, beliefs, and philosophical orientations toward health and life."

And this brings us back to the twisting influence of the philosophy. If people want to go to an alternative practitioner because they enjoy the personal attention (and I sure did), because they appreciate the holistic approach to health, because they are attracted to the underlying philosophy and are comfortable with the lack of evidence, then I say go for it. But if the field claims to be scientifically informed, as is the case with modern naturopathic medicine, then it must follow the principles of science. And practitioners should not deceive patients about what the evidence says.

Homeopathy is an easy target, for sure. It is universally accepted by those within the scientific community that it's a crock. But that is why it serves as such a damning example of the twisting power of an overriding ideological framework. If naturopathy were really evidence-based, would naturopaths provide homeopathy as a primary treatment? Is the use of homeopathy supported by "voluminous research," as claimed by the head of the British Columbia Naturopathic Association? Or do naturopaths believe in this therapy because it is part of the naturopathic tradition and accords closely with the foundational (and unscientific) principles of the field?[1]

To explore the profile of homeopathy as a naturopathic remedy, I worked with one of my students to analyze the services offered by Alberta naturopaths. We wanted to get a sense of the degree to which the most common naturopathic medicines were science-based. We examined all the websites we could find for naturopathic clinics in the province. As it turned out, the remedies most commonly offered were homeopathic. Of the 53 websites, 50 of them mentioned homeopathy. Ninety-four percent of Alberta naturopaths, therefore, explicitly advertised homeopathy, an unproven and scientifically implausible therapy that has been likened to witchcraft, as a core service. And many of the same websites tell the public that naturopathic services are evidence- and/or science-based.

Our analysis also revealed that Alberta naturopaths were offering a wide range of other more or less bizarre services, including colon cleansing. Colon cleansing is really nothing but a fancy enema. The process clears out your lower colon and is performed, or so the theory goes, in order to "detoxify" your system. There is no evidence to support the idea of detoxification (of any kind) or the use of colon cleansing. Indeed, the procedure can be both unhealthy, because it messes with your body's natural bacterial fauna, and dangerous, because it can lead to the perforation of the lower intestine. Do not get your colon cleansed—it's perfectly capable of doing the job on its own. If you see the words "cleanse" or "detoxify" used to market or explain a health product or diet, you can assume it is a scam. Sorry, Gwyneth Paltrow. Many experts I talked to thought it was such an absurd idea that it was not even worth discussing.

It makes you wonder, though. Given the lack of evidence and the existence of real risks, why do naturopaths market colon cleansing? I can only speculate, but I suspect it is the naturopathic philosophy at work again. Colon cleansing is viewed as a natural remedy (why irrigating your bowel is considered "natural" is not entirely clear to me) and it is associated with the hydrotherapy procedures first proposed by Kneipp. Clearly, the philosophical spin has a long reach and an enduring influence.

Colon cleansing also has ties to the religious and spiritual beginnings of naturopathic medicine. What is the connection? Apparently God has a clean colon and She wants you to have one too. As noted on the website for Corinthians Naturopathic College, a school that claims to be accredited by something called the American Alternative Medical Association and the American Association of Drugless Practitioners, the Bible tells us, in Corinthians 17:1, to "cleanse ourselves from all the filthiness of the flesh and spirit." And you can go to this college to learn to do just that. You can take courses such as Biblical Nutrition 101 and Cleansing and Detoxification 102 and get a Certificate in Colon Hydrotherapy in order to "help restore man in God's image," meaning, no fecal matter in the bowel.

To be as fair as possible to naturopaths, the Corinthians Naturopathic College is not one of the seven schools accredited by the Association of Accredited Naturopathic Medical Colleges (AANMC), the entity that establishes the standards necessary to allow graduates to get a licence in the jurisdictions where naturopathic medicine is a regulated health profession. (In North America there is a great deal of variation in regulatory policy. Some jurisdictions, such as British Columbia and Washington State, promote and license practitioners; others are silent or have evolving rules.) Ontario's Canadian College of Naturopathic Medicine (CCNM) is one of the seven. In fact, the CCNM is considered to be one of the world's leading and most sophisticated naturopathic education institutions. Its website states that it seeks to promote a "culture of research" to help advance "our understanding of complementary and naturopathic medicine."

So, does this top naturopathic education and research institution accept the scientific evidence that rejects remedies such as homeopathy? On the contrary, its website confirms that homeopathy is a key naturopathic treatment. It also makes the completely unscientific claims that "homeopathy restores the body to homeostasis, or healthy balance, which is considered its natural state" and that "homeopathic remedies are designed to stimulate this internal curative process." The website asserts, despite all evidence to the contrary, that "homeopathic remedies are particularly effective for depression, anxiety, allergies, infections, gynecological concerns, skin conditions, digestive problems, chronic and acute conditions including colds and flu." I suppose you could argue that homeopathic remedies are effective in the same way placebos are effective, but I don't think that's what they're trying to say. CCNM, naturally, offers classes in homeopathy.

It is worth highlighting that a stamp of approval by the Association of Accredited Naturopathic Medical Colleges is no reason to think an educational institution embraces science. The AANMC website, for example, recently featured a glowing profile of an alumnus, a naturopathic oncologist named Daniel Rubin. In the profile, which is structured like an interview, he is asked what the biggest challenge to his work is. His response, in part: "One of the greatest challenges we [naturopaths] face is the widespread public belief in the scientific method. Medicine cannot create success exclusively through clinical trials. We're too reliant on the scientific method, and it stands in our way of forging ahead." Remember, this is on the website of the association for the most respected naturopathic schools and is a statement made by a man who uses naturopathic remedies to treat cancer.

It is impossible to reconcile this kind of talk with the assertion, so often made by the advocates of alternative medicine seeking mainstream legitimization, that it is an evidence-based profession. Naturopathy is the antithesis of a science-based approach to health. (Indeed, as aptly noted by the apparently respected naturopath Daniel Rubin, the belief in the scientific method stands in the way of naturopathic medicine.) Naturopaths might be involved with research and sit on policy committees that consider evidence—a point made on several of the naturopathic websites I visited—but that does not make the field evidence-based. It is founded on a mystical and quasi-spiritual belief in a vitalistic, nature-cure life force. And any evidence that conflicts with this faith-based vision must be twisted to fit the philosophical foundation. Otherwise the very existence of the profession would be called into question. Clearly, that cannot be allowed to happen, so they keep on trying to have it both ways.

I want to re-emphasize that I respect an individual's right to believe in this Yoda-ish view of the universe. Go forth, young Jedi, and enjoy the care of your naturopath. But we should not cloud reality by calling it science-based or evidence-based health care. There is an intellectual chasm between conventional and alternative medicine. Some commentators have suggested that the gap runs so deep that it can never and should never be bridged, despite the current push to integrate the two practices.

For example, Bruce G. Charlton, a professor of theoretical medicine at the University of Buckingham and a reader in evolutionary psychiatry at Newcastle University, believes that the crucial and defining difference between alternative and conventional medicine is that alternative approaches have non-scientific explanations that are based on spiritual, mystical, or, as he put it, "intuitively appealing insights." Conventional medicine has (or seeks) biological explanations. There is a profound philosophical difference.

Charlton has written extensively on the limitations of both conventional and alternative medical practices. In fact, he told me that the issues associated with alternative medicine are a "side show" compared with growing problems with how science is done in relation to conventional medicine (a point I will tackle later in this chapter). To Charlton, alternative medicine is nothing more than a branch of New Age spirituality, "a subjectivist expression of secular modernity," as he told me in a recent correspondence. "It makes no sense," he says, "to try and integrate a subjectively based system like alternative medicine with an objective system like [conventional] medicine."

The meridians of acupuncture, for example, have no literal scientific or biological significance. Rather, Charlton argued, they are "suggestive poetic symbols of the way that life can be experienced as flows of energy." Looked at this way, acupuncture becomes an extension of a world view. It becomes a way to embrace an ideology and, to some degree, a different cultural approach to health and wellness. For those who have spiritual New Age tendencies, such an approach to health, from a psychological perspective, can play an important role in the healing process. It can provide comfort and a sense of control, for example. The same could be said about naturopathy, chiropractic, and homeopathy. They are all based on theories that have mystical or magic-like qualities: naturopathy's nature cures; chiropractic's view that vertebral subluxations can cause a range of diseases; and homeopathy's belief that like cures like. All deeply unscientific. All devoid of solid, supportive, scientifically derived evidence. But then alternative medicine, as Edzard Ernst and many of the other experts I talked with said, is, at the core, more akin to religion than science. Evidence really doesn't enter into it.

While working on this book I talked to many friends and colleagues about the therapies I was investigating. I was astounded by the number of people who thought homeopathy was an effective treatment. These were not necessarily people who used homeopathy or went to alternative practitioners. These people had simply not heard anything contradictory. This got me to thinking: given the non-evidence-based reality of many alternative therapies, why is there not a more visible critique? Are they simply drowned out by the advocates of alternative medicine?

There are, of course, many forces that create this phenomenon. For example, as with other topics covered in this book, the media plays an important role. The popular press is a primary source of health information for the general public. And, for a host of reasons, the popular press has a decidedly gentle approach to alternative medicine. Ernst noted this during our discussion in Exeter. He told me that, for some alternative therapies, the media presents a "nearly 100 percent positive spin." This is not mere speculation. His group published a study in 2006 that analyzed more than 300 British newspaper articles. It found that the coverage was largely uncritical and increasingly promotional in tone, and that most of the therapies described were not supported by research.

An Australian study published in 2008 examined more than 200 newspaper stories on alternative medicine. The researchers gave each article a score based on a variety of criteria, such as a reference to evidence, the use of an independent source of information, and mention of risks and costs. The researchers found a huge variation in the quality of the articles, ranging from a grade of 33 percent to a high of 54 percent. Some featured headlines such as "Microwave Your Flab Goodbye" and "Acupuncture Linked to IVF Success." While articles on remedies that might have a biological foundation, such as herbs and supplements, were handled better than the majority (this group got the just barely passing grade of 54 percent), most failed miserably. The authors concluded that "currently, it appears that much of the information the public receives about CAM [complementary and alternative medicine] is inaccurate or incomplete."

Research on this point has also been done in Canada. A study led by Laura Weeks, an academic who did her PhD on this issue, looked at more than 900 articles on alternative cancer treatments in newspapers and magazines. Given that cancer patients consistently identify the mass media as a primary information source to support their decisions to use complementary and alternative medicine, Weeks and her colleagues at the University of Calgary viewed this analysis as critically important. They found that alternative medicine was typically presented in a positive fashion, that it was often described as a potential cure for cancer, and that most of the articles contained little or no information on the risks and costs.

In recent correspondence about the role played by the media, Weeks told me that "a relevant factor is the need for simplicity in a necessarily short media story. Things need to be black and white, good or bad, and very straightforward in order to compose a coherent media story." Weeks believes that reporters rarely have the opportunity or time to access the appropriate experts; thus an incomplete picture is presented.

The research by scholars like Weeks paints a disappointing picture of the news media in this context, but the data did not surprise me. I am often amazed by the number of cheery, lifestyle-oriented articles in reputable newspapers that promote alternative therapies as though they were efficacious options that have resulted from rigorous scientific research. Not long ago, for example, I read this headline on the front page of my hometown newspaper, the *Edmonton Journal:* "Patients Left Bruised, but Grateful: Say Chiropractic Tool Technique Brings Relief." The article was about something called the Graston technique, which involves the use of a metal instrument to do deep massage of damaged muscles and tendons. The idea was that the "micro-trauma" caused by the massage will help the healing process. The article presented the technique as effectual; there was not a single word questioning its value. The only sources of evidence provided were quotes from happy patients and a chiropractor. (Naturally, this chiropractor markets the service on his website.) One patient provided this glowing testimonial: "The pain of the treatment is temporary, the relief long-lasting and a better option than the cortisone shots the doctor offered." The article was written by a respected local health reporter.

The *only* message the public could possibly glean from this newspaper story was that this technique has merit. It was not presented as an experimental procedure. It was presented as something that works, as an exciting new innovation.

But what did the evidence actually say? I challenged my research team, four bright individuals with considerable research skills, to find a single solid research study involving human subjects that supported the use of the Graston technique. They found nothing. Not a shred of positive evidence. In fact, most of the studies they uncovered questioned the value of the entire technique of deep massage as a remedy for damaged muscles and tendons. A summary of the research (or, more accurately, the lack of research) on the topic, posted on the website Science Based Medicine, an entity devoted to evidence-based health care, backed up my own team's findings. The author of the summary found no solid evidence for the efficacy of the Graston technique. She found several mouse studies—one pro and the others con—a pilot study, and several vaguely scientific case reports. One of the negative mouse studies came to this conclusion: instrument-assisted cross-fibre massage (e.g., the Graston technique) "had minimal effect on the final outcome of healing" as compared with other massage techniques. The one pilot study (the designation "pilot study" is generally used to describe small studies that are meant to test whether future, more robust research is justified; results from pilot studies should never be considered definitive) simply concluded that the technique worked as well as another form of massage. My skeptical side feels compelled to point out that this pilot study was published in the *Journal of Manipulative and Physiological Therapeutics,* a publication which, as its website declares, is "dedicated to the advancement of chiropractic health care."

And yet there it was, in the newspaper, a story about an unproven technique used by an alternative field of medicine. And it was on the front page. And it was all positive. How does this happen? And why does the media so frequently put a positive spin on stories about alternative medicine?

"There are many reasons," Edzard Ernst told me back in Exeter. His assessment of journalists, which is informed by both research and his frequent interaction with the news media, is as harsh as his assessment of the remedies. "The so-called health journalists often don't know the area. They often can't spell 'complementary medicine.' And the good journalists often tell me that their editors want a positive spin. People want to hear that it works."

Laura Weeks's research confirms this latter point, though she interprets it as a systemic problem. She concluded that the information in the media about alternative medicine was "insufficient to assist patients with informed decision-making" and that the messages conform to the "commercial interests of media outlets, as coverage appears to be focused around entertainment rather than information provision."

The media does not make money educating the public about the realities of health remedies. They are an entertainment-based industry. The truth about alternative remedies does not sell papers as well as a happy message does.

"In the news business," Jodie Sinnema told me, "there is always pressure to produce quirky or unique stories, especially when there isn't much hard news happening ... and editors love the unusual." Sinnema, a respected journalist, has produced numerous thoroughly researched and comprehensive stories on complex health issues, and was a runner-up at the 2010 National Newspaper Awards for best health reporting. She was also the author of the Graston technique article I read on the front page of the *Edmonton Journal*. The NNA honour was not for her article on the Graston technique. "I considered the story a soft feature—one that would be better suited to the living versus news section, where the stories are less hard-hitting or critical," Sinnema recalled. Her experience as a health reporter has also taught her that "it's very difficult to get accurate scientific data about alternative techniques."Of course, the news media is not the only source of information about alternative medicine. The public increasingly relies on the internet for all kinds of news. Unfortunately, research has consistently shown that websites are a poor resource. One analysis of 150 websites undertaken by a team from the University of Texas found that 25 percent "contained statements that could lead to direct physical harm if acted upon." Ernst's group at Exeter has conducted several similar studies. His 2008 analysis concluded that "many sources of information contain advice on therapies that have no evidence base and may put patients at risk."

When it comes to alternative medicine, the twist is everywhere.

The waves were bad on the second to last night of the cruise. It was a leg of our journey that put us far from the protection of the coastal islands, and the open sea was causing the big ship to sway with considerable force. At dinner that night the wine sloshed, the rolls rolled, and the kids turned green.

Just before the meal was served I saw the face of one of my nephews go still. He stared blankly over his dinner plate. Then, with his mouth hanging open, he did the telltale half-gag-vomit motion. This was the prelude to the explosion of puke every parent at the table knew was about to arrive. And it had the effect of a starting gun on all who saw it. As the go-to vomit expert, I moved the quickest. Instinctively and instantaneously I hopped from the table, grabbed the young, pale-green seafarer and dashed to the closest washroom. His dad ran ahead to hold open doors and clear the way. We arrived at a porcelain receptacle with milliseconds to spare.

That was when it hit me. I was holding a puking kid, but I wasn't puking. My delicate stomach was withstanding the twin threat of a heaving ocean and a heaving kid. A few days before, on the fourth day of the cruise, I'd been forced to place a pharmaceutical product behind my ear, something called a scopolamine patch. This product released a steady stream of an antihistamine drug through my skin. And it seemed to be very effective.

So I had passed the vomit threshold and motion sickness had won the battle. Pharmaceuticals had been deployed. But, ironically, it didn't happen on the ship. It wasn't waves that brought me to my knees and forced me to ingest various commercially produced chemical solutions. I will provide more on my descent (actually, if I'm going to be accurate, it was more of an ascent) into nausea hell later. On *this* night I did not vomit.

But despite the patch, the intense motion of the ship was still enough to bring on a dull sensation of nausea. I decided to head back to my cabin early. I lay on my bed and utilized the acupressure techniques suggested by the acupuncturist. And, to my surprise, the nausea seemed to subside. Perhaps it was wishful thinking. Perhaps it was simply the fact that I was lying down. Perhaps it was a placebo effect facilitated by my knowledge that, in fact, there is evidence to support the use of acupressure as a tonic for nausea. Or perhaps this basic manoeuvre actually worked.

As discussed earlier, acupuncture is a remedy that has long played an important role in traditional Chinese medicine. A recent article in the *New England Journal of Medicine* explained that acupuncture is based on "an ancient physiological system (not based on Western scientific empiricism) in which health is seen as the result of harmony among bodily functions and between body and nature." Acupuncture is used to fix the internal disharmony that blocks the flow of the body's vital energy, known as *qi*.

From a scientific perspective this is all nonsense. A vital life energy is no more real than the equally ancient idea that the movement of the planets controls our fate. But despite the fact that the underlying justification is a mystical, quasi-religious philosophy, there are some studies that *suggest* acupuncture can be helpful for things like pain control and nausea.

The fact that acupuncture or acupressure could have an impact on the body is not entirely surprising. Unlike homeopathy, which clearly has no effect beyond the inducement of a placebo effect and has no active agents with the potential to confer benefit, acupuncture involves sticking a needle into the skin. This physical act could, theoretically, trigger a biological response that does *something*. Studies have found, for example, that acupuncture can cause the release of neurotransmitters. It is conceivable that this biological reaction to the needles or acupressure can relieve some symptoms. It is also conceivable that this was the physiological mechanism that helped to quell my queasiness on that wavy night.

A rigorously executed 2009 analysis of all available, methodologically robust clinical trials (40 research studies were included) found that stimulation of the P6 acupoint, which is a spot on the inner arm about four centimetres down from the wrist joint, can help to reduce nausea and vomiting and that it does this with few side effects. While this analysis was focused on the use of acupressure to help with post-surgery nausea, it is fair to say that there is a growing consensus that acupressure can help with nausea more generally.

So, despite my skepticism, I am not *against* alternative medicine. But I am against ignoring what the available evidence tells us. I am against not recognizing the vested interest that alternative practitioners have in continuing to propagate unscientific myths as science. I am against misinforming the public. And I am against pretending alternative medicine is something it is not. We need to study and learn what we can from and about alternative medicine, including gaining an understanding of why people go to alternative practitioners and what is missing from the therapeutic experience in the context of conventional medicine.

As I see it, we should do two things: accept that most alternative medicines are not, at their core, a science-informed health-care practice and are more akin to a myth-based belief system (which is fine and may allow alternative medicine to play an important, non-medical role in the healing process) and strive to examine them in a dispassionate and objective manner in order to determine, from a scientific perspective, if and how they work. If we take the former approach, we should not pretend they are evidence-based and neither should their practitioners, funders, regulators, and our health-care systems. If we adopt the latter approach, and I think we should, then we must accept the outcomes of good research.[2] Remedies that don't work, don't work. Remedies that are shown to be effective should not be viewed as "alternative medicine" but simply as "medicine."

In order to disentangle the genuine therapeutic effects from the false promises we need to develop research methods that allow us to strip away the forces that distort the truth about the effectiveness of remedies. When it comes to many alternative remedies, particularly an approach like acupuncture, this is not easy. In addition to the philosophical spin and the placebo effect, there are the potential benefits derived from interaction with the practitioner. Both my naturopath and my acupuncturist impressed me with their professional and caring approach. This kind of personal attention can enhance the placebo effect and may add to the therapeutic benefit.

A study published in 2010 highlights how complex this kind of effect can be. The study examined the use of acupuncture to relieve the pain experienced by people with osteoarthritis. This is a painful and common condition that affects millions of people in North America, and acupuncture is an often-recommended treatment. Indeed, even Ernst has endorsed it. In a 2008 article he outlined the few alternative practices that are worth funding because they "did more good than harm." There aren't many—a "meager twenty," as he says in the article—and most are some form of herbal remedy for a very specific condition. In the article Ernst argued that there are only two uses of acupuncture that pass the good-enough-to-consider-funding threshold: acupuncture for nausea and for osteoarthritis pain (though he thinks conventional options are probably more effective for both conditions). In other words, this is a use of acupuncture that even some of the skeptics are comfortable supporting.

The 2010 study, however, casts doubt on the idea that there is a biological mechanism involved, rather than a placebo effect. This study, which was conducted by a team that included experts in acupuncture, involved an eloquent methodology that was specifically designed to get to the role of practitioner-induced placebo effect. Researchers randomly assigned more than 500 patients to a range of treatment arms. All the patients—save those in the control arm who got no acupuncture treatment—received a 30-minute assessment by an expert in Chinese medicine (not unlike the process I underwent) before undergoing the acupuncture. For half the patients the therapists used high-expectation language such as "I think this will work for you" and "I've had a lot of success with treating knee pain." Therapists treating the other half used a more neutral language. Both groups were then randomized to receive either real acupuncture or sham treatment. The patients, of course, did not know which treatment they were receiving.

The results were illuminating: acupuncture as practised by experts in Chinese medicine is *not* superior to sham acupuncture.

Those patients who received fake acupuncture—meaning that the needles were stuck at a shallow depth in spots that had nothing to do with meridians—did as well as those who received a treatment that carefully followed the rules of Chinese medicine. More importantly, the study found that the communication style of the practitioner had an impact on the perception of how well the treatment worked. Those patients who heard high-expectation language perceived greater pain reduction. As the authors concluded, this study suggested that "the perceived benefits of acupuncture may be partially mediated through placebo effects related to the acupuncturists' behavior."

Here is my skeptical (but I believe realistic), bottom-line approach to the assessment of alternative remedies: assume that all claims of actual, non-placebo-effect benefits are false. Assume that the information you are being given about alternative remedies, at least by alternative providers, the media, and websites, is incomplete or inaccurate. If you take this approach, your assessment of the remedy will likely be correct 89 percent of the time. (I will shortly explain the seemingly overprecise 89 percent.) I am going to generously guesstimate that one remedy in ten (10 percent) is backed by research that suggests a slightly-better-than-placebo benefit, even though the mechanism of action is unclear. These are the remedies, such as acupuncture for nausea and osteoporosis pain, that seem to provide a degree of scientifically supportable but still uncertain health benefit.

I have three important caveats to add to this guesstimate. First, the therapeutic effect is usually mild and temporary. These are not strong effects. There are *no* miracle cures. Second, much of the relevant research is still incomplete or in its early stages. As noted in the acupuncture study above, it is very difficult to tease out the degree to which the placebo effect is producing the therapeutic response. More high-quality research is needed. Once this work is done, I suspect we will find that the placebo effect plays a significant role in many of the remedies shown to have some beneficial effect. Third, and most important, even if an alternative remedy works, it is *not* confirmation of the underlying, pseudoscientific philosophy (meaning, the fact that the P6 acupressure works is not proof of the existence of a life-force energy flow through a system of meridians). A finding that one modality is somewhat efficacious does not make the entire alternative approach "scientific."

The remaining 1 percent in my guesstimate is for those few remedies that actually seem to make a real physiological difference. These are the alternative medicines that should be considered, simply, medicine. They are the remedies that are supported by enough research to demonstrate that the benefits are not merely placebo-derived. A good example of this kind of remedy is the use of St. John's wort, a herbal remedy, for depression. Studies have consistently shown that this herb works as well as standard antidepressants and has fewer side effects. It is, therefore, an evidence-based medicine, not an *alternative* medicine.

There are, then, alternative remedies that research has shown to be truly effective. But, as noted in the paper by Ernst that highlights the few alternative medicines that work, there are still not many that can be said to fall into the category of truly effective. I stand by my guesstimate. In fact, 89 percent might be a bit generous.

In the preceding pages I have, admittedly, skated over the surface of a socially and scientifically complex phenomenon. Much has been left out. I have not, for example, discussed the possible harm associated with the marketing of alternative medicines, such as when homeopathic vaccines are recommended instead of real, life-saving ones, or when patients take alternative remedies that can be dangerous (such as kava), or when they undergo treatments that can adversely interact with conventional remedies. I have not considered the myriad forces driving people to alternative providers, such as lack of access to primary-care physicians. And I have looked at only a handful of alternative approaches. Still, I believe the overarching thesis regarding the twisting influence of the alternative philosophies is applicable to the entire area.

That said, there is one important critique of *my* critique of alternative medicine that I have not yet tackled. It's a critique I have heard for years, one that goes like this: "Okay, it might be the case that the truth about alternative medicine gets twisted, but the promise of conventional medicine gets twisted worse!" My response is always the same: correct. But that in no way invalidates my assessment of alternative remedies. It merely highlights that there are also profound problems with the science and messaging associated with conventional medicine. That's a whole other shelf's worth of books.

A consideration of the myriad issues associated with pharmaceuticals takes me back to the cruise.

One morning while still at sea, the Caulfield clan awoke to some disturbing news. When I joined the family for breakfast I saw worried faces and heard much medical hypothesizing. My sister-in-law Catherine was having serious difficulty with one eye. The iris was completely dilated and would not respond to light. As the family discussed the scary development, Catherine stood in silence wearing dark sunglasses and a distressed expression. Her three young boys were eating Fruit Loops at a nearby table.

There were nine adults in this family group, three of whom were doctors, one a nurse, and one a deluded academic convinced he had an informed and valuable opinion on every topic (guess who?). Needless to say there were a lot of theories about what was going on. The fact that only one of her eyes was dilated was especially troublesome. This hinted that something was interfering with the nerve. The scariest possibilities: brain tumour or aneurysm. Only Catherine, a doctor, was willing to voice these possible diagnoses, but everyone else was thinking about them. We focused on practical issues. Would she have to get off the ship?

Where could she get an MRI in the middle of the Alaskan wilderness? Did Sarah Palin have an MRI machine in her basement?

We all remained anxious for a good part of the day until my wife, Joanne, also a doctor, had an idea. She went back to our cabin and checked out the side effects associated with the scopolamine patch. And there, in tiny writing on the side of the box, was a warning that the drug could cause eye problems and that you should wash your hands after attaching and removing the medicated adhesive. Catherine had recently removed a patch and obviously had rubbed her eye. Crisis resolved.

At this point in the trip I had still not thrown up, so I was sticking with my alternative approaches to motion sickness. I was hardly bright-eyed and bushy-tailed, but I was still much further from the vomit threshold than I had anticipated. And Catherine's experience with scopolamine gave me another reason to avoid pharmaceuticals. While homeopathy and acupressure may have questionable value, they also have no chance of producing the kind of reactions that remind you that you're mortal. I was beginning to think I would make it through our journey without ingesting any drugs. It was starting to look like a qualified victory for alternative remedies.

But it was not to be.

The next day our itinerary took us into the old mining town of Skagway. Early in the morning our colossal ship docked behind another colossal ship. The two liners towered over the few streets that make up the entirety of the town. We all marched down a tiny gangplank and headed to the historic train station built in 1898 to support the Yukon gold rush. Our entire vacation crew, all nine adults and ten offspring, piled into a passenger car on a wonderfully preserved train. As the train pulled out of the station, passing a graveyard full of unsuccessful gunslingers and the shells of poorly preserved locomotives, our car began to sway.

At first I wasn't worried. Normally, trains are one of my preferred modes of transport. Though the smooth click-clacking ride often lulls me to sleep, it rarely gives me motion sickness. But this old Skagway train was not the kind I usually rode. It was not London's Heathrow Express or Tokyo's Narita bullet train, but rather an antique diesel-powered engine and cars chugging up the side of a mountain on rails that were built to haul hardened gold prospectors, not a gastrically challenged academic. Comfort and smoothness of ride were likely low on the priority list of nineteenth-century engineers. "I think I can, I think I can, I think I can," I said to myself as waves of nausea began to wash over me. It was no use. Less than 45 minutes after we left the station I was in the train's little bathroom. I had lost the battle. The vomit threshold had been crossed before the train reached the peak of the mountain.

Joanne noticed my predicament, and knowing the rules of my remedies game she handed me a pharmaceutical she had been saving for this very moment, which she seemed to know was inevitable. I took two of the little pink pills, and within 20 minutes I was in the kind of slumber that produces drool, grunts, and startled surprise when abruptly ended. The sleep was another drug-induced side effect, but this time it was one I welcomed.

When I got back to the ship, I carefully placed a scopolamine patch behind my right ear (and washed my hands). And from that moment on my motion sickness problems were largely resolved.

I wrote earlier that I agreed with the statement that the twisting forces that characterize conventional medicine are worse than in the area of alternative medicine. I believe this to be true. But I don't hold this opinion because I think conventional remedies don't work. Nor do I hold this opinion because conventional medicine is bound to some quasi-religious philosophy that clouds our perspective. No, the misrepresentation of conventional medicine is different from that which afflicts alternative therapies, which mostly don't work and are bound to a particular world view.

I think the forces that distort conventional medicine are worse, or even more reprehensible, because conventional medicine is *supposed* to be evidence-based. Given the lazy thinking underpinning alternative medicine, I expect the truth about practices such as homeopathy to be perverted. But in the evidence-based world of conventional medicine, the application of the principles of science—the rigorous and dispassionate collection of data—is meant to allow us to avoid or at least dampen any obfuscating forces. That is one of the core values of science. But despite conventional medicine's commitment to science, the truth still gets twisted, and often with disastrous results.

In 2010, the US National Institutes of Health (NIH), the largest source of public funding for biomedical research in the world, had a budget of approximately $32 billion. That sounds like a lot of money. But this pot must support a significant portion of all biomedical research in the United States, which is the country that spends more than any other on research. And getting NIH funding is not easy. It is an extremely competitive process involving the submission of a long and detailed grant application, the kind of application that can take an entire summer to write (I have had more than one "laptop holiday"). The success rate is around 20 percent, and that's for some of the world's greatest scientists. Of the nearly 50,000 researchers applying for funds in 2009—and these are, in general, medical doctors and scientists affiliated with a research university—four out of five went away empty handed. The Canadian Institutes of Health Research (CIHR), Canada's primary biomedical research entity, has a much smaller budget, around $1 billion per year. But the success rate for applicants is about the same as in the United States. It's a brutally competitive business just to secure the funds for your biomedical research, let alone to then actually *do* the research.

By comparison, in 2007, the annual revenue for the top nine US pharmaceutical companies exceeded $150 billion. The recent downturn in the world economy has reduced profits, but, in 2009, they still sat at around $50 billion. That's just for the top earners. If you add up the money made by the entire industry, the figure would increase by many billions. Pharmaceuticals has long been one of the most profitable industries in the world. In fact, according to *Fortune* magazine's listing of the 500 most profitable companies, pharmaceuticals ranked as the world's third most profitable industry in 2009 (its close cousin, medical products and equipment, ranked fourth). The Fortune 500 is densely populated with Big Pharma (the nickname given to the major pharmaceutical companies).

There is nothing wrong with making money, even lots of money. Profitability does not make an industry evil. But when there is a huge amount of money in play, market forces are intense. It's only natural that the people in charge seek ways to optimize their earnings. But in the context of pharmaceuticals, this sometimes means nudging scientific data in a direction that favours the pharmaceutical industry, which inevitably means emphasizing benefits over risks, and effectiveness over limitations. As we have seen in the preceding chapters, all industries do this. But the products sold by Big Pharma are usually the result of years and years of biomedical research aimed at the accumulation of scientific data. This isn't like selling jeans or shampoo. Virtually any medication is the result of a long chain of studies, including clinical trials with real patients who may have put their health at risk in the name of science.

There are several reasons why the pharmaceuticals industry is a particularly noxious example of the mishandling of science. The first is the increasingly significant role that drugs play in the world's health-care systems. We are a society awash in pills. Drug spending represents almost 20 percent of the total health-care spending in developed countries. (The list of economically developed countries is often said to correspond to the membership of the Organization for Economic Co-operation and Development [OECD]. Currently there are 41 members, including the United States, Great Britain, Japan, France,

Germany, Canada, Spain, Poland, Mexico, Korea, and the newest member, Peru.)

That 20 percent is a massive portion of the developed world's health investment. In the United States, the country that spends the most (by far) on health care, the average per person spending on pharmaceuticals is over $700 per year. In Canada, we do a bit better, but per person costs are still remarkably high, over $500 per year. And this spending is increasing rapidly. Spending on pharmaceuticals across OECD countries has increased by an average of 32 percent since 1998. The annual combined pharmaceutical budget of the OECD countries now hovers around half a trillion dollars. Think about that. Half a trillion dollars is $500 billion. That's $500,000 million. On pills.

Given all the money devoted to purchasing drugs, much of it taxpayer money, you would hope that the science behind the drugs would be rock solid—or, at least, as solid as possible. If we, as a society, are going to allocate such a massive and measurable portion of the world's wealth to an industry founded on the application of science to health-care concerns, the science better be good.

But the biggest reason to fear the distortion of science in this context is, of course, that people's health is at stake. As highlighted by the adverse reaction my sister-in-law experienced on our cruise (and this was the side effect of a fairly common drug for a relatively minor condition), all pharmaceuticals are potentially powerful and dangerous. They are, after all, chemicals that cause real biological changes in our bodies. And they are chemicals that we, at least in the developed world, are consuming with increasing frequency.

I saw a wonderful display recently at the British Museum that illustrates the pervasive and bizarre role pharmaceuticals play in our lives. On the main floor, right behind a gigantic Easter Island head, was an art installation, *Cradle to Grave* (2003) by Susie Freeman, David Critchley, and Dr. Liz Lee, that incorporated thousands of tablets and pills into a long piece of fabric. It was a lengthy, checkerboard drug rug. At first I was put off (a pretty literal approach), but then I realized that these 14,000 pills represent the average number of prescription drugs consumed by an individual over a lifetime. The tapestry is huge, 13 metres long and half a metre wide. It is enough to make you gag.

The budgets of the NIH and the CIHR are significant, but they are dwarfed by the money generated by the pharmaceutical industry. Big Pharma can throw billions of dollars into industry-friendly research and associated marketing strategies. In fact, 75 percent of the clinical trials (studies involving patients) that get published in the top medical journals are funded by industry. This means that 75 percent of all the scientific evidence being produced in relation to drugs, one of the most important and costly components of all the world's health-care systems, is being twisted in one way or another in the pursuit of financial gain. Yes, human benefit is also often the result, but it's rarely the sole rationale for the research and investment.

So much has been written about the abuses committed by the pharmaceutical industry that I do not need to go into detail. There have been numerous well-publicized horror stories, such as the overt manipulation of research results in relation to the diabetes drug Avandia, the painkiller Vioxx, and various menopausal hormone-replacement therapies. These are calculated and sophisticated actions done in the name of profit that have, some have suggested, resulted in thousands of deaths. These are immoral acts that go well beyond the mere subtle and perhaps inadvertent twisting of facts indulged in by practitioners of alternative medicine.

These types of scandals set the tone for any discussion of the twisting influence of Big Pharma. It's now taken for granted by most in the biomedical research community and by health professional organizations that the pharmaceutical industry has a perverse impact on the presentation of the relevant science. No one can argue with a straight face that the pharmaceutical industry is not trying to influence how the world views its products. Indeed, pop culture has embraced this perspective and usually frames Big Pharma as malevolent, manipulative, and exploitive. In movies, for example, the industry has emerged as the new bad guy, right up there with terrorists and gangsters. In the 2005 Academy Award–winning movie *The Constant Gardener,* for instance, Rachel Weisz and Ralph Fiennes battle a pharmaceutical company that is not afraid to murder in the name of profit.

How does the pharmaceutical industry twist science? In every way imaginable. From the selection of the research topics to the writing of the peer-reviewed articles. From the press releases issued to the media about new drugs and diseases to the creation of the clinical practice guidelines passed on to your doctor. The close personal relationship systematically established between Big Pharma sales reps and GPs every day in every city throughout the developed world is a crucial part of this picture. If your doctor recommends a new drug, there's a good chance it's because it's been personally pitched to him or her by a Big Pharma sales rep with whom your doctor has a relationship.

Consider for a moment the influence of the industry on the way academic journals choose, review, and then publish research articles. For an academic scientist, publication in a good journal—such as the *New England Journal of Medicine,* the *Lancet,* the *Journal of the American Medical Association (JAMA),* the *Canadian Medical Association Journal,* and the *British Medical Journal*—is considered a significant career achievement. In fact, in the academic world, it is one of the single most noteworthy things you can do. Academics are weighed and measured by their publication records; it's their career currency.

In part, this prestige comes from the fact that it is tremendously difficult to get published in good journals, and, as a result, they are perceived as the keepers of the best scientific information.

Perhaps more than any other source, academic journals are meant to be a reserve of independent science, a place where the best work is shared with colleagues and distributed to professionals for consideration and criticism. The process by which manuscripts are submitted and assessed is designed to lead to the selection of papers of high quality and relevance in a manner that promotes the principles of scientific integrity. By and large, this is done through peer review.

Peer review, the practice of sending manuscripts to experts for evaluation, has been around for centuries. It is the bedrock of academic publication. In 1731 the Royal Society of Edinburgh published *Medical Essays and Observations*, which some consider the first peer-reviewed collection of medical articles. Since then, independent peer review has been adopted by all respected biomedical academic publications. The *British Medical Journal*, for example, has been using peer review since 1893. For top journals, the rejection rate is tremendously high. It is a competitive and ruthless process. *JAMA* and the *New England Journal of Medicine* turn down about 95 percent of the articles submitted to them.

So, given the history and prestige of medical publications, their well-established review process, and their high rejection rate, you'd think they would be beyond the influence of industry.

You would be wrong.

In a well-known 1998 study published in the *New England Journal of Medicine*, Henry Stelfox and colleagues explored the influence of industry funding on the results of peer-reviewed medical publications. The researchers were trying to figure out if money from pharmaceutical companies led to the publication of articles that favoured the industry. They found a strong association between authors' published positions on the safety of a particular drug and "their financial relationships with pharmaceutical manufacturers." In plain language, Big Pharma funding influences peer-reviewed results. The data is striking: among the authors of original research papers, reviews, and letters to the editor that were supportive of a drug's use, "96 percent had financial relationships with the drugs' manufacturers; for publications deemed neutral or critical the figure was only 60 percent and 37 percent respectively."

More recently, a Canadian study led by Mohit Bhandari came to a similar conclusion: "Industry funding has a significant influence" on research results. Specifically, the authors found that "industry-funded trials are more likely to be associated with statistically significant pro-industry findings, both in medical trials and surgical interventions." These results have been replicated in numerous other studies, and the conclusions are always the same: there is a systematic bias that favours the products made by the company funding the research.

How can this happen? How can the best science journals in the world consistently publish data that has a bias toward industry-funded products? How can they publish data that is so obviously influenced by the profit motive as opposed to the pursuit of objective knowledge? Are the journals in cahoots with industry? Are the scientists involved in the research deliberately skewing the results to make the funders happy?

And, the cynic in me wonders, given the amount of money involved, Why are we even shocked?

Often, the data does not tell us how, exactly, the bias toward industry happens. In some ways, this is a key point. Market forces are both powerful and subtle. They don't always work in an overt and obvious manner. (I will point out shortly, however, that there are a number of identifiable, measurable, and rather dubious industry tactics.) Nor do people always realize that they are being swayed. We can measure the influence of industry involvement, but not necessarily the how and why; it's not as if Big Pharma is ever going to step up and say, Hey, here's how we do it! But profit and personal gain can be assumed to be lurking somewhere in the background. Again, this is not an attack on capitalism, which is a key driver of innovation, but rather a question of science's ability to withstand, and defend itself against, the pressures that are placed on it *by* capitalism.

To illustrate the subtle influence of industry on behaviour, let's take a brief detour into the promotion of drug sales within the medical community. Few physicians would admit that perks and promotions from the pharmaceutical industry influence their clinical practice. But the pharmaceutical industry spends, in the United States alone, $57 billion per year on promotion, around 25 percent of their revenue. This includes $5 billion for sales representatives. The sales reps are the individuals who meet one-on-one with doctors to talk about the company's products. Incredibly, there is one drug representative for every ten doctors. If you include expenses and training, the pharmaceutical industry spends approximately $150,000 annually for each drug representative who works with general practitioners, and on average $330,000 a year for those working with specialists.

Using everything from small gifts, free drug samples, dinners, and plain old flattery, this army of sales representatives is a primary driver of drug use.[3] But despite all the money, time, and attention that are lavished upon them, studies have shown that doctors don't think they are influenced by Big Pharma. A 2003 study found that doctors have no ethical problem with the perks; as the authors of the study concluded, doctors "continue to have a rather permissive view about a variety of marketing activities."

If you are feeling courageous, the next time you see your doctor ask if her clinical decisions are influenced by sales representatives. I predict she will say no. I also predict, with only slightly less certainty, that she will not be thrilled with the question (so conduct this experiment at the end of your visit). In your doctor's heart, she probably truly believes that the thousands spent each year on *her* by the pharmaceutical industry are not having an impact on which products she is thinking of putting into your body.

But how could this possibly be true? How could every doctor be immune to influence? Research has shown that even though doctors do not think drug marketing influences them, they readily admit that they get most of their information about new drugs from sales representatives. We must be somewhat sympathetic here, because we all know that doctors are busy. They do not have a lot of time to spend perusing the latest literature on a new drug or treatment plan. My wife, Joanne, is a family physician. She is, of course, one of the brightest and most conscientious doctors in the country, if not the entire world. But even she struggles to find time to read the professional literature, as evidenced by the stack of medical journals sitting in our bedroom. Pharmaceutical representatives help to fill this information gap simply by being available. One study found that 70 percent of general practitioners regarded drug representatives as an "expedient means of acquiring and processing drug information and keeping up to date with new products."

It's no surprise, then, that other research has shown that all of this exposure to the pharmaceutical industry—exposure that, incredibly, starts in medical school—is associated with requests by doctors to add specific drugs to hospital formularies (the list of drugs available at a given hospital) and prescribing practices. After a visit by a sales representative, the rate of drug prescriptions by the doctor increases. Even small gifts can have a powerful, subconscious influence on behaviour, even when there is no explicit expectation to reciprocate. Even a gift as small as a pen or drug sample can impact the objectivity of the recipient, a behavioural response that the pharmaceutical industry counts on. If it didn't work, they'd have stopped handing out freebies decades ago.

A wide variety of other marketing tactics has been found to influence physician behaviour. A study by Ashley Wazana found that everything from attending industry-sponsored education events to watching presentations by industry-selected speakers is "associated with non-rational prescribing." A systematic review, this one from 2010, came to a similarly damning conclusion: "With rare exceptions, studies of exposure to information provided directly by pharmaceutical companies have found associations with higher prescribing frequency, higher costs, or lower prescribing quality or have not found significant associations." The authors were so concerned about the influence of the industry on physician behaviour that they "recommend that practitioners follow the precautionary principle and thus avoid exposure to information from pharmaceutical companies."

What might be even worse than the fact that Big Pharma spends tens of billions of dollars a year on promotion in North America, worse even that they know with certainty that they can influence practitioners with gifts and relationships, is that at the end of it all they aren't really providing the practitioners or the public with high-quality information. It's a double-whammy, worst-case scenario. It's a lose-lose. Systematic analysis of pharmaceutical advertising has found that it is frequently misleading. For example, one study reported that 57 percent of the advertisements had no educational value and 44 percent could lead to improper prescribing. Overall, there really isn't any way to characterize the entire situation other than to say that there's an odds-on chance that your doctor's clinical practice is being influenced by bad information that favours the financial bottom line of the pharmaceutical industry.

Let's return to the more fundamental twist: the sad reality that even the science that is meant to be objective, the work published in independent peer-reviewed journals, is being adulterated. How do the pharmaceutical companies influence what gets into the world's leading medical journals? Among the most audacious and, arguably, the most unethical strategies they use are ghost-writing and guest authorship.

You would think that the authors listed on a research paper wrote the article and conducted or contributed to the research.

But, as it turns out, this isn't always the case. The pharmaceutical industry often pays academic researchers to "author" papers largely written by others. In this context, I am using the term "author" quite loosely, as the "author's" involvement is likely minimal, non-existent or, at best, unclear. This activity has been called "guest authorship" and it is done to give the research the sheen of academic credibility and independence. A paper written solely by researchers employed by a pharmaceutical company might look suspicious, especially if the paper is touting the benefits of the company's drug. But get a few independent academics to become "authors" and the study acquires instant credibility.

A ghostwriter is a person (or company) that writes a paper but is not credited with authorship. The use of ghostwriters allows the industry to retain control of the message about a product. Indeed, ghostwriting is often part of a complex and comprehensive publication planning process utilized by pharmaceutical companies, often over years, to emphasize the benefits of a drug and minimize chatter about harms or limitations.

Ghostwriting and guest authorship are obvious and blatant departures from academic norms: international publication guidelines generally state that authors must make a substantial contribution to the study and the writing of the paper. Despite this, recent research has found that ghostwriting and the use of guest authors is relatively common. For example, a 2007 analysis of 44 peer-reviewed pharmaceutical studies revealed that 40 of them had some contribution from a ghostwriter, usually an industry statistician. Ninety percent of all studies, in other words, were not independently written by the person whose name was under the title.

Guest authorship also appears to be widespread, occurring in 16 to 26 percent of all articles in medical journals. One of the most notorious examples involved the major pharmaceutical company Merck and Co. In September 2004 this industry giant was forced to pull one of its top-selling products, a pain medication called

Vioxx, off the shelves. This action was taken because emerging research suggested that the drug increased the risk of heart attack and stroke. There is evidence that Merck knew about the danger and wilfully suppressed the data. This revelation led to worldwide media attention and significant legal action (billboards advertising law firms looking for Vioxx plaintiffs peppered the American landscape: "Get a Free Vioxx Claim Case Evaluation!"). Eventually, Merck agreed to pay $4.85 billion (yes, billion) to settle the bulk of the claims.

Because of the litigation, many of the internal company documents associated with Vioxx research were made public, including letters to authors, draft research papers, and email correspondence. These materials gave observers a unique opportunity to look into the publication strategies of a major pharmaceutical company, an opportunity seized by four scholars: Joseph Ross from Mount Sinai School of Medicine, Kevin Hill from Harvard Medical School, David Egilman from Brown University, and Harlan Krumholz from Yale. In a groundbreaking paper published in *JAMA* in April 2008, Ross and his colleagues provided a comprehensive analysis of Merck and Co.'s approach to the study of Vioxx. Their conclusions are shocking, particularly when one considers the fact that the company's spin on the scientific data about the risks and benefits of Vioxx may have contributed to the loss of lives.

Ross and his team unearthed clear evidence of inappropriate guest authorship. Merck had an established practice of finding trustworthy academic scientists to list as "authors," often lead authors, *after* the relevant research was completed and the bulk of the manuscripts were written. Ross and his colleagues even included in their article images of a final, peer-reviewed study and of the rough draft of the same study. The draft paper, an investigation into the effects of Vioxx on patients with Alzheimer's, listed a number of scientists from the Merck Research Laboratories as authors, but in the space that ought to have listed the first author, the most prestigious position, was the phrase "External Author?" In the final, peer-reviewed publication, three new scientists are listed, including two from prestigious academic institutions: the first author is Leon Thai from the University of California, the second Steven Ferris from New York University's School of Medicine, and the third Louis Kirby from Pivotal Research Center in Arizona.

Given that the draft, which contains no reference whatsoever to the three individuals who were eventually listed as the lead authors, was nearly identical to the final publication, Ross and colleagues took this to be more than a little suggestive of questionable publication practices. They wrote: "Although there are minor differences in language and organization between the draft and final versions of the manuscript (particularly in the abstract, as opposed to the text), the results presented are almost identical, reinforcing that the trial itself and the analyses were complete before the academically affiliated investigators were involved in the manuscript."

The Vioxx litigation documents revealed numerous other examples of guest authorship. Ross and his colleagues found evidence of ghostwriting that was appallingly systematic. There was even email correspondence from the company, Scientific Therapeutics Information, employed by Merck to write the research papers (which is to say, Merck's ghostwriters). These messages provided Merck with an "update on the development and estimated delivery dates for various publications related to VIOXX." As part of the update, the ghostwriters supplied a nicely organized status report, including information on the "intended author" and where the paper should be submitted, that is, the intended peer-reviewed journal. A private for-profit firm, in other words, was clearly orchestrating the authorship and placement of what the external world would view as a wholly independent piece of research.

Ross and his colleagues provided a revealing, and deeply depressing, glimpse into the world of ghostwriting and guest authorship, concluding that "Merck used a systematic strategy to facilitate the publication of guest authored and ghostwritten medical literature." Naturally, this strategy is used by Merck and its competitors to control the production of scientific data so as to maximize profit.[4] Given that profit is the understood goal of the industry, it should be no surprise that they would use every technique (even these ethically questionable techniques) at their disposal to bring the money in. From the perspective of the pharmaceutical industry, this is the purpose of peer-reviewed publications, a point highlighted in a header from a 2000 Pfizer sales document. "What is the purpose of publications?" the document's headline ran. The answer: "[The] purpose of data is to support, directly or indirectly, the marketing of our product."

Pharmaceuticals is a big, profit-motivated industry, and it will inevitably act as such.

But why would respected academics get mixed up in this practice? Unfortunately, we have no definitive answers; we can only speculate. Indeed, in a 2008 *Globe and Mail* article on the Vioxx scandal, it is reported that despite the documented evidence, Steven Ferris continued to deny that he was merely a paid "guest author." We should not discount the possibility that the scientists who become "guest authors" believe, for whatever reason, that it is appropriate for them to be listed as a lead author on a study they clearly did not lead. Perhaps the practice has become so widespread that it feels normal and acceptable. Perhaps it's because guest authors often get an honorarium. (The Ross study of the Vioxx case notes that the fees for academics who agreed to be guest authors on papers ghostwritten by medical publishing companies ranged from $750 to $2500. Not a huge amount of money, but not bad for doing little more than providing your name.)[5] Another significant reason could very well be the prestige associated with being a lead author on a peer-reviewed publication in a major journal. The pressure to publish is an important part of academic life. When "publish or perish" is the prevailing state of mind, it cannot help but be tempting to have a near-finished paper plopped down on your desk. You avoid all that boring, tedious, and time-consuming work normally associated with lead authorship, such as the development of a hypothesis, designing a study, attracting research funds, recruiting patients, analyzing data, and writing up results. All you have to do is read it, perhaps make a few suggestions about the wording, and sign off. Not the most reassuring picture of scientific inquiry, I admit, but you can be sure it happens.

Of course, ghostwriting and guest authorship are not the only strategies employed by the pharmaceutical industry to influence what the peer-reviewed research says about its products. There is ample evidence that the pharmaceutical industry carefully selects which studies to submit for publication, and (surprise!) there is a strong bias toward conclusions that favour the industry. In addition, industry-funded studies are designed in a manner that makes the sponsor's drug look good. For example, a study might compare a sponsor's product with that made by a competitor, but the competitor's drug will be administered at an inadequate dose—and of course it won't work as well. And there is also evidence that studies which seem to be coming to conclusions that do not favour the sponsor's product are brought to a halt. The premature termination of a clinical trial is a surefire way to avoid the publication of a negative result.

What we have here is an academic train wreck. The goal of scientific inquiry, remember, is to uncover objective facts about the world. In the context of pharmaceutical research, those facts are meant to lead to the production and use of new drugs to save and improve lives. To this end, billions of dollars are spent on clinical research trials. Billions more are spent by individuals and health systems to purchase the products of that research. But despite this massive investment in science, the pressure to make a profit constantly obscures and twists the scientific results. Science is under attack, from so many corners, that sometimes it seems futile to put up a defence. But it's worth defending. In some ways, that's the core message of this book. *Science, when done properly, is worth defending.* And it's worth defending because when it's not twisted, it actually can make us healthier. We just need to know how to recognize the twist.

"That would have been a good ending to your book," my brother said.

He was referring to the dramatic spell of gagging that was triggered by my consumption of an overdose quantity of homeopathic sleeping pills. I tried to gulp down dozens in one go and I nearly choked on the wad of sugary white tablets. Given my negative depiction of alternative remedies, death by ultra-dilution would have been a fitting end—a homeopathic revenge that could hardly be considered a mere placebo effect.

Why was I swallowing homeopathic pills like they were M&Ms? This performance was the final act of an experiment I conducted on my friends and family. It was a re-enactment of the UK protest where hundreds of people attempted to overdose on homeopathic remedies. For my experiment, however, I didn't recruit willing participants. Instead, I sprung the idea on each person. They didn't know my overdose offer was coming until I put the pills in front of them.

A few days before my choking fit, I went to a local natural food store and bought a bunch of homeopathic remedies. I wanted remedies that, if they worked at all, would produce some kind of noticeable physiological response. Pills for sleeping and anxiety seemed a good choice.

With my alternative remedies in hand, I made a surprise homeopathic house call at my brother-in-law's place. Dave is a respected orthopaedic surgeon and a pretty rational guy. I explained what current evidence says about homeopathic remedies and told him about the UK protest. "So, Dave, how about a few sugar tablets? Can I convince you to take a bunch of homeopathic sleeping pills?" I asked.

"I don't think so," he said after a bit of a pause.

"Why not?"

"I'm not ready for bed."

Janine, Dave's wife, is a nurse. She listened to my pitch. Would she swallow a few? Nope. "I'm just not sure what's in them."

"*Nothing* is in them," I told her, a bit frustrated. Still, she would not budge.

Next, I tried my neighbour, a pathologist. "Ummmm … I don't think so."

How about my wife, Joanne? "No thanks. I'm a fairly cautious person."

My sister-in-law Catherine, a family physician, agreed to take two anti-anxiety pills, "but only because I'm feeling anxious," she explained.

My brother, an artist, was willing to gobble a bunch. Ditto his wife, Akiko, also an artist. Perhaps they were hoping for a homeopathic, psychedelically inspiring experience.

But, in the end, I could not get any of my physician friends and family members to throw down a few sugar pills. Between them they have had decades of advanced biomedical education. During their careers they have had to evaluate evidence and make recommendations to their patients about thousands of different pharmaceutical products and medical procedures. Still, even though they all said they doubted the efficacy of homeopathic remedies, they remained cautious. The twist persists.

I had promised them all that I would also consume mega-doses of the homeopathic solutions. So, right after a family dinner, I stood at our kitchen counter and swallowed as many of the pills as I could. Hence the choking fit. The dosage recommendation for the homeopathic insomnia remedy was two to three pills near bedtime. I consumed almost the entire bottle, dozens of pills. I also ate all the anti-anxiety tablets.

Hours later I didn't feel any less anxious than usual, but I did find that dealing with the unrelenting nature of the twisting influences in the world of health remedies was a bit exhausting. I headed to bed early.

5

MAGIC
SIMPLICITY AND THE UNTWISTED TRUTH

For the year that I worked on this book I completely submerged myself in the world of health science. I interviewed an army of experts. And I got personally engaged in every topic I covered. I exercised like a maniac, went on an ultra-healthy diet, got my genes tested, and tried a variety of remedies. It was a fascinating journey that led to some surprising conclusions about what the science actually says and the degree to which this message gets twisted. I also discovered some new and, in general, ego-deflating things about myself. I am far from the genetically ideal athlete that I had always imagined myself to be. I was (and likely will be again) a bit pudgier than I imagined. And, despite decades of dedication to exercise, it turns out that I can barely handle the fitness routine of a wimpy Hollywood actor. In sum, I am a sluggish, shortish softy with a physique-destroying love of chocolate-covered peanuts.

But other than this personal reality check, what are the take-away lessons from this journey?

On the one hand, the results of my research point to a disheartening conclusion, which is, basically, that nothing works. Despite the immense diet, fitness, and remedy industries, very little actually does what it promises to do. A scan of your genes will not tell you what will happen in your future; for most of us, it's no more useful than the numbers you get from a weigh scale or a blood pressure cuff. It is nearly impossible to transform your body through exercise alone. You cannot get sexy abs instantly or even after weeks of intense work. There is no such thing as toning, and virtually every fitness gimmick is just that, a gimmick. To lose weight you have to eat fewer calories than you burn. Sadly, we don't need many calories. There is no shortcut to weight loss. And even if you can take off the pounds, keeping them off is the real challenge (a challenge I'm facing as I write these words). The failure rate is so high that some experts I interviewed thought that sustained weight loss is ... sigh ... impossible.

Finally, most of the remedies offered by alternative practitioners work no better than a placebo, and the pharmaceutical industry has such a tight grip on the production of the relevant science that it is difficult to trust any available information about any drug, whether it comes from your physician, a medical journal, or an advertisement.

In short, there are no magical answers. This should not come as a surprise. If it were easy, we would all be healthy. If alternative therapies worked, we would have verifiable data demonstrating their efficacy. If losing weight and getting fit could be attained by utilizing a metabolism-enhancing, colon-cleansing yoga move, we would all be slim, cut, and have pristine innards. Alas, this is not the world we live in.

On the other hand, there is another way to look at the results of this inquiry. This is the glass-half-full view. If you want to optimize your health, the steps are, in fact, surprisingly simple. The steps are not *easy*—real effort is required—but they are straightforward. It isn't complicated.

This is a liberating realization. It means you can shut out most of the noise. Ignore the advertisements. Ignore the miracle-cure promises made by alternative practitioners. Ignore any marketing message that includes the words *detoxify, cleanse, metabolism, enhance, boost, energize, vitalize,* or *revitalize.* Ignore the hype!

Don't get fooled by the images of sexy abs that are such a huge part of Western culture. Don't worry about the genetic predispositions that have been handed to you in the biological lottery of life. Unless you have one of the rare single-gene disorders, like cystic fibrosis, or one of the relatively uncommon, highly predictive mutations, genetic information is simply not that valuable. Don't get suckered into buying useless potions and practices that are wrapped inside an ideologically fuzzy and truth-obscuring blanket. They will only empty your wallet. And don't get too excited when the media report on some big health breakthrough, especially if the story is based on a single study. True breakthroughs are rare. Think of science as a slow and iterative process. As geneticist Jim Evans told us, science is a slog. Two steps forward, one and a half steps back.

What, then, are the straightforward steps to maximum health? First, exercise often and with intensity (intervals work best) and include some resistance training. Second, eat small portion sizes, no junk food, and make sure 50 percent of what goes in your mouth is a real fruit or vegetable. Third, try your best to maintain a healthy weight (yes, this is insanely tough—but we should, at least, try). Fourth, do not smoke, and drink only moderate amounts of alcohol. And fifth, take all the well-known and simple injury-prevention measures, such as wearing a seatbelt in the car and a bicycle helmet when you go riding.

Once you cut through the twisted messages that saturate our world, you will find that all the available evidence tells us that these five actions are, by far, the key elements of a healthy lifestyle. One expert I corresponded with for the diet chapter, Walter Willett from Harvard, figures that healthy food choices, physical activity, and not smoking could prevent over 80 percent of coronary heart disease, 70 percent of strokes, and 90 percent of type 2 diabetes.

There are other measures, such as getting a good night's sleep, that are important, and future research might compel me to add them to the list. And we should be conscious of eating certain other foods in addition to fruits and vegetables, like fish, berries, and whole grains. Also, there are things that should probably be avoided, such as sodium and trans fats. But, in the big picture, these five actions remain essential. All the other stuff you hear about—such as the alleged importance of various supplements and the craze for organic foods—will likely have only a marginal impact on *individual* health. If you are not doing one of the big five, worrying about the details—such as a slightly increased genetic predisposition to some common disease or the cleanliness of your colon—is ridiculous.

There are no magical cures or programs. But the simplicity of the untwisted truth has an almost magical quality.

Throughout this book I have sought to do two things: first, to highlight the ways in which the science of health is distorted, and, second, to provide advice, based on the best available evidence, about how to achieve a healthy state. However, I haven't said much about what can be done to avoid the twist. Given the significance of the health sciences to contemporary society, addressing this issue should be a social priority.

Academics and policy-makers throughout the world have pondered this question. What can be done to stop the twist? A comprehensive review of all the solutions that have been suggested is beyond the scope of this book. But before I wrap up it is worth considering one overarching solution, and it can be summed up in one word: independence.

"Adam Smith would be furious," Jerome Reichman told me as we stood before a statue of the famous Scottish political philosopher. Reichman looked up at the statue, shook his head, and continued: "It's hard to believe all the ways they mess up the research, including circulating wrong information and wilfully suppressing important data."

The "they" Reichman is referring to is the pharmaceutical industry. Reichman and I were both speaking at an event at the University of Edinburgh. I couldn't believe my good fortune. I had recently read an article that he co-authored with colleagues from Duke University. Meeting him in Edinburgh provided an opportunity to reflect on his paper on the home turf of the world's most famous capitalist theorist. In the article, Reichman, who is a law professor, argues for the creation of a completely independent entity to conduct all clinical trials. In other words, he thinks that the research needs to be taken out of the hands of industry. Specifically, he proposes that the "federal government should oversee and manage both the process of drug testing and the dissemination of test results." The Duke group argue that the "direct link between the clinical trial sponsor" (i.e., the pharmaceutical industry) and the drug tester must be broken; otherwise, the science will always be "sub-optimal from the standpoint of public health and safety."

Reichman's team is not the only group to recommend this radical approach. Indeed, Marcia Angell, the former editor-in-chief of the *New England Journal of Medicine,* has called for the creation of a similar institute within the existing framework of the US National Institutes of Health. And several Canadian scholars have made similar recommendations. In 2008, two philosophers, Mathieu Doucet and Sergio Sismondo, argued that all the existing solutions to the problems created by industry involvement in research—such as requiring a declaration of conflict of interest, instituting mandatory reporting of negative research results, and trying to limit physician exposure to the drug industry—are bound to fail. The market is simply too powerful. The only way to avoid this distortion of science is to take control of the production of knowledge away from those who profit from it.

Reichman couldn't agree more. "The priorities are simply different. They can never be the same. The priority of the drug companies is to make money, not to find big health benefits. They want money. Period." He told me this in a tone that signalled his sense of futility. There is simply too much money at stake for the drug companies to ever let control slip from their hands. But

Reichman emphasizes that the desire to control the research can *only* be for nefarious motives. The centralization of knowledge production would, in theory, benefit all. It would result in higher-quality data. It would stop the hiding of negative trial results, the use of ghost authorship, and the selective publication of favourable results. It would be a far less costly system because resources could be pooled. There would be an economy of scale. And it would result in data that everyone could trust, including health-care providers, administrators, and the public.

"It's economics 101. It would be vastly more efficient," Reichman said. "There would be less wasteful duplication of studies and the most needed clinical research could get priority. It would also be easier to recruit patients as research subjects. Why not have an independent entity produce the data? It's better for everyone. Unless, of course, you want to control and manipulate the data for the benefit of profit, which the drug companies can't admit, of course."

And this is why Reichman thinks the current situation would infuriate Adam Smith, the individual whose name is most often thrown around by those espousing the virtues of the market. By allowing market forces to twist science, we are allowing the power of profit to ruin something that gains its social value by virtue of its dispassion and objectivity. Profit is an incentivizing force. It is not about a dispassionate and objective assessment of the world. It is about generating a profit-enhancing perspective, regardless of the truth. Science is one of the few tools that can counter this propensity. Indeed, Adam Smith noted that good science acts as a tonic to the warping effects of unbridled zeal. Specifically, he said: "Science is the great antidote to the poison of enthusiasm and superstition."

While a radical restructuring of the way in which pharmaceutical research is done seems unlikely (though we must not retreat from this battle), we can all learn from the spirit of the recommendations made by Reichman and others. Information that comes from an independent source—one that is removed from the profit motive, ideological spin, or other warping influences—is almost always the most valuable and trustworthy. Independence is a key tool in the fight against the forces that distort.

You can use this fact to your benefit in your own assessment of health information. Whenever possible, seek out independent sources of information. For example, a project like the Cochrane Collaboration is a marvellous resource. This project, which I used often while researching this book, provides systematic reviews of a wide range of remedies and preventative strategies. The reviews use the best available information, are written by top scholars, and take into consideration conflicts of interest that may impact the validity of relevant data. The reviews contain a lay summary of the major conclusions, and so are accessible to all.

There are many other useful sources, and, thanks to the internet, much of this information is now available to the general public. For example, summaries of literature put out by expert committees, like the UK Parliamentary Committee that analyzed the value of homeopathy, and respected scientific organizations, especially if they are at arm's length from a controlling entity, can be tremendously rich sources of information.

Of course, I don't need to reiterate that independence is important. Research has consistently shown that the public trusts sources of information that are seen to be independent. A large survey of both Canadians and Americans found that university researchers funded by public research grants were among the most trusted voices as compared with just about anyone else, such as the media, industry, NGOs, and politicians. Researchers funded by industry, even if they are housed in universities, were among the least trusted. The authors of the study found that "most people rest their assessment of credibility on the degree to which the person or institution is perceived to be at arm's length and independent of controlling and/or funding influencers." People are not dumb. They know an overt twisting influence when they see one. And, as we have seen, nothing twists like money.

But we also need to recognize that there is a lack of independence, so to speak, when we are making assessments about our own status, including our height, weight, calorie consumption, response to remedies, and chances of obtaining sexy abs. Indeed, personal experience can be awfully deceiving. It is one of the most powerful twisting influences. It tells us that the world is flat, the sun revolves around the Earth, and homeopathy works. In the context of health, relying on personal experience alone can be especially problematic as our perceptions are filtered through a thick fog made up of our desires, values, hopes, dreams, exposure to marketing messages, and evolutionarily pre-programmed predispositions (i.e., a hankering for sex and survival). The scientific method is specifically designed to counter this kind of observational bias. That is why, for example, you may think that a particular remedy works but a large, scientifically rigorous study says that it doesn't.

In the end, the most consistently reliable information remains the data that comes from scientists working within universities or other reasonably independent research institutions.[1] Yes, the most respected science journals and researchers are subject to twisting forces, even when there is no overt commercial presence associated with the research. As we saw in the genetics chapter, scientific enthusiasms and the need to attract and sustain research funding can also adversely impact the way in which research is represented.

A fascinating paper, published in *PLoS Medicine* in 2008, highlights the degree to which this is so. The authors of the paper used economic principles to analyze the value of published data. They concluded that due to the intense pressure placed on academics to publish frequently and in the most prestigious journals, there is an incentive to exaggerate the significance of their findings. There's also pressure on the journals to preferentially select the papers with the most spectacular results, thus encouraging and disseminating the hype. In economic terms this incentive is called the "winner's curse."

I met Neal Young, the lead author of this provocative paper, in his office at the National Institutes of Health in Bethesda, Maryland. Young is a principal investigator in hematology at the National Heart, Lung, and Blood Institute. Writing an economics-based critique of the publication process is not really his day job. But he became fascinated with the way in which research is presented to the public.

"There always seems to be a degree of hype," Young told me. "Very little of what is promised, from a health benefit perspective, ever materializes. I was curious why these top papers in top journals get it wrong so often.

"It is clear that the current system, which puts so much pressure on researchers to get career-establishing publications in top journals, forces exaggeration. The winner's curse skews the data. Less than 1 percent of the studies that promise results make it to the clinic. Perhaps less."

But despite the problem of the winner's curse, Young believes in the scientific process. "Most academics take the publication process—peer review—very seriously," he said. "Some don't, for sure. Some of the scientists who are asked to review papers for publications may be motivated by vanity or jealousy. But most in the scientific community take it seriously. This helps science to remain self-correcting. The truth eventually emerges. It may take longer than is ideal, but the process usually works."

When I read Young's paper I knew it would provide the perfect conclusion to my journey through the world of health science. It touches on so many themes in this book, including the fact that there are forces that distort science in almost every realm of the knowledge-production universe. But Young's paper is also an affirmation of the unique value of scientific knowledge. Rather than a condemnation of science, it is a reminder of how precious it is. We must be constantly attentive to the potential for twisting forces to strip science of its worth.

Young's paper ends with a quote from Karl Popper, the famous philosopher of science: "Knowledge and the search for truth are still the strongest motives of scientific discovery." As a society we must do our best to ensure that this remains so. If we lose independent and trustworthy scientific inquiry, we lose our only way to counter the increasingly powerful forces that twist the truth about our health.

So, I leave you with my simple tools, the cure, for untangling the twist: be skeptical, be scientific, be self-aware, be patient, and look for the best, most independent information.

The day I finished writing this book was the same day our family celebrated my youngest boy's seventh birthday. The day involved hot dogs, laser tag, chips, bowling, and a world-class homemade cake, complete with electric-green icing depicting a 1950s-style sci-fi rocket ship and spaceman. I enjoyed it all. It was a magical day.

NOTES

CHAPTER 1: FITNESS

1. A related social force that can distort the truth is a phenomenon called "white hat bias." There is evidence of a bias toward publishing and reporting on research that seems to support some righteous idea. A 2010 paper published in the *International Journal of Obesity* found that when the subject matter of the study is presumed to be a social good (e.g., breastfeeding) or a social bad (e.g., sugar-sweetened drinks) there is, at times, a bias in "the presentation of research literature to other scientists and to the public at large, a bias sufficient to misguide readers." In other words, the data is twisted to support a position or policy that is presumed to be noble.
2. Even if it hasn't permeated the popular consciousness, this "dose response" reality (i.e., the more time spent being active and the more intense the activity, the better) has been acknowledged in the most recent physical fitness guidelines, including Gary O'Donovan's consensus paper and recommendations from the US and Canadian governments. For example, the 2008 Physical Activity Guidelines for Americans, produced by the Department of Health and Human Services (www.health.gov/paguidelines/guidelines/summary.aspx), recommends that all adults do "at least 150 minutes (2 hours and 30 minutes) a week of moderate-intensity, or 75 minutes (1 hour and 15

minutes) a week of vigorous-intensity aerobic physical activity" and "should also do muscle-strengthening activities that are moderate or high intensity and involve all major muscle groups on 2 or more days a week." The Guidelines produced by the Public Health Agency of Canada have a similar "dose response" approach (see www.phac-aspc.gc.ca/hp-ps/hl-mvs/pag-gap/downloads-eng.php). Two reviews have confirmed the health benefits of this approach—both published in 2010 in the *International Journal of Behavioral Nutrition and Physical Activity.*

3. I want to emphasize that the definition of "work hard" will vary depending on the fitness level of the individual. For some people, it may simply mean varying between a vigorous and strolling walking pace. For the very fit, it will likely require intense sprints. The key is to challenge your system and force adaptation. Of course, you should consult a physician before attempting this kind of routine if you have any reason to be concerned about intense exercise.

CHAPTER 2: DIET

1. I do not, for example, say much on the important topics of sodium (see www.cdc.gov/salt) or trans fats (see www.cdc.gov/nutrition/everyone/basics/fat/transfat.html). But I will say this: avoid both.
2. In some ways, the DXA is simply a more accurate predictor of the health risks associated with excess weight than the BMI, which is the traditional measure. The BMI, or body mass index, is only a rough measure of body composition. The BMI is a ratio of height and weight. If you have a solid or muscular frame, you may have a high BMI but little body fat. For example, Olympic gold medallist and former world record holder Maurice Greene is listed at 5'9" (which, considering height inflation, likely means 5'8.5") and at around 180 pounds. I saw him run in the 2001 World Championships. At that time, it looked like the man had zero body fat. But his BMI would categorize him as overweight. Similarly, a tall and pudgy individual with less than an ideal amount of lean muscle could have a healthy

BMI. There are also limitations with the use of BMI in the prediction of good clinical outcomes. A recent University of Alberta study found that BMI was not a great predictor of recovery times for heart patients. The study author, Antigone Oreopoulos, told me that "more muscle, which can mean a higher BMI, is associated with favourable outcomes." (She also noted that this finding emphasizes the importance of resistance training, a point noted in the last chapter.) That said, as a way to get a rough approximation the BMI remains a valuable tool. But from a personal perspective, Antigone thinks that "waist circumference might be the best measure. It's tough to lose weight, but if your waist size is coming down, you're making good progress."

3. In some regions, there are restrictions on advertising aimed at children. In Quebec, for example, the provincial Consumer Protection Act forbids advertising products that are designed specifically for children under 13. In 1989, the Supreme Court of Canada upheld the law, noting: "Television advertising intended for children is inherently manipulative. It aims at promoting products by convincing those who are always ready to believe anything."
4. A bit of this weight was lost pre–fat test (DXA), as I started the "no poison food" before I got to the DXA laboratory. Of course, this means that my pre-DXA fat percentage was likely even higher! I am putting my "official" pre-experiment weight at 198 pounds. I thought this was 198 pounds of pure muscle. Nope. Not even close.
5. You should also question the accuracy of the calorie counts provided by restaurants and commercially prepared foods, especially for reduced-calorie items. At least one study found the offered information to be consistently inaccurate. The study, published in 2010 in the *Journal of the American Dietary Association,* found that some restaurant reduced-calorie offerings had 200 percent more calories than advertised. Given this kind of research, it is safe to assume that you are eating more calories than the label says you are eating. I use a 10–20 percent bump.
6. No surprise; this is a common reaction. A 2011 study from the US

found that people who are on a diet are angrier than individuals who are not trying to exert self-control.

7. The way people metabolize food does differ. And genetics plays a role. There are undoubtedly a few people who are outliers, individuals who either need lots more food than most or much less. But current research tells us that the vast majority of adults fall within the 1800–2200 calorie range. Recent population genetic studies have shown that having a genetic predisposition toward obesity—including carrying the FTO gene, the most predictive of the obesity-related genes found to date—likely accounts for only three to seven extra pounds.
8. Coincidentally (and it really was a coincidence), as I was finishing the writing of this book, the US government switched from the familiar food pyramid approach to providing dietary advice to a plate approach that is very similar to what I suggest above. See ChooseMyPlate.gov.
9. The supplement story is, indeed, complex. There are some supplements, as exemplified by the current push to consume vitamin D and fish oil, which are beneficial. Others, such as vitamin E, could be harmful. Some populations, like pregnant women and the elderly, have different nutritional needs, and supplements are often recommended (e.g., folic acid and iron supplements for pregnant women). But, in general, if you are eating a healthy diet, you don't need to take a bunch of supplements. As nicely summarized in the 2010 Dietary Guidelines for Americans, we should simply aim to meet our "nutrient requirements through a healthy eating pattern that includes nutrient-dense forms of foods, while balancing calorie intake with energy expenditure." And later, "Dietary supplements are recommended only for specific population subgroups or in specific situations."
10. Despite my skeptical conclusions about keeping weight off, I am an optimist. People like Diane Finegood are an inspiration. It can be done. I'm going to do my best to keep it off. And there is research that tells us what successful dieters do to *keep* it off. Here are the

main themes (you will note much overlap with the points made in this chapter): eat small portions, eat a healthy and filling breakfast, weigh yourself (very) often, drink only water (fruit juice and pop are diet killers), eat lots of fruits and vegetables, keep a weight-loss diary, get a good night's sleep (studies have consistently shown that poor sleep makes weight loss and maintenance tough), do your best to learn the calorie content of food, and be lucky enough to have a great support system and family. The other thing that successful dieters do is exercise. Now, I know this might sound like it contradicts one of the main conclusions of the last chapter, but it does not. Exercise did not make the weight come off, but it seems to be correlated with keeping it off. Numerous studies have found this to be true. It may be that exercise is simply correlated with a healthier lifestyle. Or, perhaps, it allows just enough of a modest calorie burn to make maintenance easier.

CHAPTER 4: REMEDIES

1. It should be noted that advocates of homeopathy often claim that you can't use traditional scientific methods to study homeopathy because the treatments must be tailored to respond to individual health needs (even though you can buy them as pre-made tablets, just like pharmaceuticals). But, as noted by Ernst and others, you can design studies to explore almost anything. During our interview he told me that he's never heard a convincing argument about why, if given the appropriate sample size and research design, you cannot study homeopathy. After 200 years of research, if the homeopathy was even marginally effective, there would be at least some supportive data. But, as noted by entities like the British Parliament's Science and Technology Committee, that data does not exist. To combat the scientific implausibility of the ultra-dilutions method, homeopaths have rolled out the field of quantum physics—suggesting that the quantum properties of homeopathy can be physical without being observable. It is a strange paradox that a complex and revolutionary field of scientific inquiry, quantum physics, is being used to explain

something that is the antithesis of science, homeopathy. Of course, few in the scientific community take this explanation seriously. The advocates of homeopathy are trying every kind of argument—often contradictory—to fend off the "pseudo-science" label. Some, like the head of the BC Naturopathic Association, claim that it is supported by traditional scientific research. Others claim that you can't study the effects of homeopathy. And still others say that it's supported by science but that you can't see it work because it follows the principles of quantum mechanics. Desperate arguments all. This scrambling defence of homeopathy highlights the power of the philosophical spin.

2. There is an ongoing debate about the degree to which public research funds should be devoted to researching the efficacy of alternative remedies. Critics claim that many of these remedies are so preposterous, with no possible scientifically sound mechanism of action, that awarding grants to study the area is a waste of valuable resources that could be spent on more plausible and potentially beneficial research. Further, the critics claim, alternative-medicine researchers should be required to justify their work within the framework of the rules that apply to all areas of health-science research. Supporters note that the popularity of alternative therapies dictates the need for scientific investigation. The establishment of the federally funded National Center for Complementary and Alternative Medicine (NCCAM) in the US has been a focus of much of this debate. As noted in a 2009 *Washington Post* article on the increasing pressure to shut down the NCCAM, "The notion that the world's best-known medical research agency sponsors studies of homeopathy, acupuncture, therapeutic touch, and herbal medicine has always rankled many scientists."
3. The gift-for-doctors practice is now viewed as so problematic that some jurisdictions, such as Massachusetts, have passed laws to greatly restrict the kind of perks pharmaceutical representatives can give to physicians. As a result of this kind of policy move (and efforts by the industry to self-regulate), recent research suggests that physician exposure to pharmaceutical-industry pressure might be

decreasing. But there seems little doubt that it remains a profound ethical issue.

4. There are many other examples of this kind of data manipulation. A 2010 paper by Georgetown University researcher Adriane J. Fugh-Berman meticulously outlines the numerous and systematic ways in which ghostwriting was used in the context of hormone replacement therapy. The paper, which was published in the journal *PLoS Medicine,* found that there were dozens of ghostwritten reviews and commentaries published in peer-reviewed medical journals with the goal of promoting the unproven benefits of menopausal hormone replacement therapy. These publications were also used to minimize the harm of menopausal hormone therapy. For example, the pharmaceutical company Wyeth used ghostwritten articles to mitigate the perceived risks of breast cancer and to promote the "unsupported cardiovascular 'benefits' of HT, and to promote off-label, unproven uses of HT such as the prevention of dementia, Parkinson's disease, vision problems, and wrinkles." It is interesting to note that, despite all evidence to the contrary, many physicians still believe that the benefits of hormone replacement therapy outweigh the risks. As noted by Fugh-Berman, this perception is likely the legacy of years of carefully crafted and misleading "research" articles—thus demonstrating the disturbing effectiveness of ghostwriting and the planned publication strategy.
5. The Ross analysis did not provide details of how much compensation, if any, Thai, Ferris, and Kirby received.

CHAPTER 5: MAGIC

1. There is increasing pressure on university researchers to commercialize their work. In other words, governments and funders expect university researchers to come up with products that will create wealth. University research is seen not as a source of knowledge, but as a source of revenue. This trend, which in countries like Canada is a relatively recent phenomenon, threatens to erode public trust in the most important source of independent scientific information.

SOURCES

In the text of the book I did my best to provide enough identifying information about every study, policy, or website I used so that it can be found in the following list of references. I also included publications that are not mentioned explicitly in the text but that were important to my work. For some of the references I briefly state how it is used or why it is relevant.

CHAPTER 1: FITNESS

"2000 Physical Activity Monitor." Canadian Fitness and Lifestyle Research Institute (2000). www.cflri.ca/eng/provincial_data/pam2000/canada.php.

Adams, Stephen. "Fitness Fanatics 'Have High Sex Drives.'" *Telegraph* 14 October 2010.

Akuthota, Venu, et al. "Core Stability Exercise Principles." *Current Sports Medicine Reports* 7.1 (2008): 39–44.

Alleyne, Richard. "Yoga Protects the Brain from Depression: Practising Yoga Really Does Relax Your Mind as Well as Your Body More Than Other Types of Exercise, a New Study Claims." *Telegraph* 20 August 2010.

Anders, Mark. "Does Yoga Really Do the Body Good?" *ACE FitnessMatters* (September/October 2005): 7–9.

Anderson, Fiona. "Vancouver-Based Lululemon's Profits Triple: Best First Quarter Rewards Vancouver-Based Seller of Workout Gear." *Vancouver Sun* 11 June 2010.

Bea, Jennifer W., et al. "Resistance Training Predicts 6-yr Body Composition Change in Postmenopausal Women." *Medicine and Science in Sports and Exercise* 42.7 (2010): 1286–95.

Begley, Sharon. "Forget Crossword Puzzles—Study Says 3 Hours of Exercise a Week Can Bolster Memory, Intellect." *Wall Street Journal Online* 16 November 2006. http://online.wsj.com/article_email/SB116364034566424589-lMyQjAxMDE2NjEzNjYxNDYwWj.html.

Behringer, Michael, et al. "Effects of Resistance Training in Children and Adolescents: A Meta-Analysis." *Pediatrics* 126.5 (November 2010): e1199–e1210.

Billard, Mary. "Yoga Manifesto." *New York Times* 25 April 2010.

Bilski, Jan, et al. "Effects of Exercise on Appetite and Food Intake Regulation." *Medicina Sportiva* 13.2 (2009): 82–94.

Brimelow, Adam. "Link between Inactivity and Obesity Queried." BBC News 8 July 2010. www.bbc.co.uk/news/10545542. Summary of study suggesting that inactivity is not the cause of obesity.

Brownell, Kelly D., and Kenneth E. Warner. "The Perils of Ignoring History: Big Tobacco Played Dirty and Millions Died. How Similar Is Big Food?" *The Milbank Quarterly* 87.1 (2009): 259–94.

Burgomaster, Kirsten A., et al. "Similar Metabolic Adaptations during Exercise after Low Volume Sprint Interval and Traditional Endurance Training in Humans." *Journal of Physiology* 586.1 (2008): 151–60. Benefits of 30-second intervals.

Canadian Press. "Big Weights Not Needed to Build Brawn: Study." MSN Canada 11 August 2010. This article states that you need to push yourself when lifting but don't necessarily need to lift a huge amount of weight.

Chalé-Rush, Angela, and Roger A. Fielding. "Relative Importance of Aerobic versus Resistance Training for Healthy Aging." *Current Cardiovascular Risk Reports* 2 (2008): 311–17.

Cheng, Maria. "No Time to Exercise? No Problem. Intense Interval

Training Could Slash Hours Off Your Workout." Associated Press 25 February 2010. Quote from Helgerud about value of interval training.

Cochran, Heidi Lynn. "Shaping a Form: The Evolution of the Feminine Ideal." *Intersections* 7 (2009): 41–55.

Cope, M.B., and D.B. Allison. "White Hat Bias: Examples of Its Presence in Obesity Research and a Call for Renewed Commitment to Faithfulness in Research Reporting." *International Journal of Obesity* 34 (2010): 84–88.

Daley, Amanda J., et al. "Randomized Trial of Exercise Therapy in Women Treated for Breast Cancer." *Journal of Clinical Oncology* 25.13 (2007): 1713–21.

Di Cagno, Alessandra, et al. "Preexercise Static Stretching Effect on Leaping Performance in Elite Rhythmic Gymnasts." *Journal of Strength and Conditioning Research* 24 (2010): 1995–2000. Study suggests that deep stretching before leaping performance may negatively affect rhythmic gymnastics judges' evaluations.

Donaldson, Liam. "On the State of the Public Health." *2009 Annual Report of the Chief Medical Officer*. UK Department of Health, 2009. www.dh.gov.uk/en/Publicationsandstatistics/Publications/Annual Reports/DH_113912.

"Eight Out of 10 Women Would Have Surgery to Lose Weight Rather Than through Diet and Exercise." *Telegraph* 11 March 2010.

"Fact Sheet—New Physical Activity Recommendations." Canadian Society for Exercise Physiology (CSEP) and ParticipACTION (2010). www.participaction.com/ecms.ashx/PressReleases/en/CSEP_PAC-FactSheet-FINAL-EN.pdf.

Fell, James S. "The Myth of Ripped Muscles and Calorie Burns." *L.A. Times* 16 May 2011. www.latimes.com/health/la-he-fitness-muscle-myth-20110516,0,7183828,print.story. Comments on the myth that building muscle will increase your metabolism and allow you to eat more.

Gerber, Markus, and Uwe Pühse. "Review Article: Do Exercise and Fitness Protect against Stress-Induced Health Complaints? A Review of the Literature." *Scandinavian Journal of Public Health* 37 (2009): 801–19.

"Girls 'Smoke to Stay Thin.'" BBC News 23 April 2003. http://news.bbc.co.uk/2/hi/health/2969757.stm.

Gorman, Kim. "Physical Exertion during Activity and Exercise." *Obesity and Weight Management* 5.5 (2009): 242–43.

Grogan, Sarah, et al. "Smoking to Stay Thin or Giving Up to Save Face? Young Men and Women Talk about Appearance Concerns and Smoking." *British Journal of Health Psychology* 14 (2009): 175–86.

Hardman, Adrianne E. "Physical Activity and Cancer Risk." *Proceedings of the Nutrition Society* 60 (2001): 107–13.

Harris, M. "It Takes More Than a Year to Truly Get into Exercising: Study." *Vancouver Sun* 10 September 2010. Reports on the length of time it takes to get into a fitness routine.

Hawley, John A. "Specificity of Training Adaptation: Time for a Rethink?" *Journal of Physiology* 586 (2008): 1–2.

Herbert, Ron D., and Michael Gabriel. "Effects of Stretching before and after Exercise on Muscle Soreness and Risk of Injury: Systematic Review." *British Medical Journal* 325 (2002): 468.

Honjo, Kaori, and Michael Siegel. "Perceived Importance of Being Thin and Smoking Initiation among Young Girls." *Tobacco Control* 12 (2003): 289–95.

"How Much Physical Activity Do Adults Need?" Centers for Disease Control and Prevention (2008). www.cdc.gov/physicalactivity/everyone/guidelines/adults.html.

Jamtvedt, Gro, et al. "A Pragmatic Randomised Trial of Stretching before and after Physical Activity to Prevent Injury and Soreness." *British Journal of Sports Medicine* 44 (2010): 1002–9.

Kelleher, Andrew R., et al. "The Metabolic Costs of Reciprocal Supersets vs. Traditional Resistance Exercise in Young Recreationally Active Adults." *Journal of Strength and Conditioning Research* 24.4 (2010): 1043–51.

Kelley, George A., and Kristi S. Kelley. "Impact of Progressive Resistance Training on Lipids and Lipoproteins in Adults: A Meta-Analysis of

Randomized Controlled Trials." *Preventive Medicine* 48.1 (2009): 9–19.

Kesäniemi, Antero, et al. "Advancing the Future of Physical Activity Guidelines in Canada: An Independent Expert Panel Interpretation of the Evidence." *International Journal of Behavioral Nutrition and Physical Activity* 7 (2010): 41. Review of the Canadian guidelines and the benefits of exercise.

Kirk, Sara F.L., Tarra L. Penney, and Yoni Freedhoff. "Running Away with the Facts on Food and Fatness." *Public Health Nutrition* 13.1 (2010): 147–48.

Lawton, Julia, et al. "'I Can't Do Any Serious Exercise': Barriers to Physical Activity amongst People of Pakistani and Indian Origin with Type 2 Diabetes." *Health Education Research* 21.1 (2006): 43–54.

Lee, I-Min. "Physical Activity and Weight Gain Prevention." *Journal of the American Medical Association* 303.12 (2010): 1173–79.

Little, Jonathan P., et al. "A Practical Model of Low-Volume High-Intensity Interval Training Induces Mitochondrial Biogenesis in Human Skeletal Muscle: Potential Mechanisms." *Journal of Physiology* 588 (2010): 1011–22.

"Lululemon Stretches Q2 Profit." CBC News 10 September 2010. Company profits. www.cbc.ca/news/business/story/2010/09/10/lululemon-second-quarter-profit.html.

Lund University. "High-Intensity Interval Training Is Time-Efficient and Effective, Study Suggests." *ScienceDaily* 12 March 2010. www.sciencedaily.com/releases/2010/03/100311123639.htm.

Macfarlane, D.J., and G.N. Thomas. "Exercise and Diet in Weight Management: Updating What Works." *British Journal of Sports Medicine* 44 (2010): 1197–1201. A useful summary of recent research showing the difficulty of using exercise alone to lose weight, the importance of intensity, and the modest benefits of moderate-intensity activities like walking.

Macy, Dayna. "Yoga Journal Releases 2008 'Yoga in America' Market Study: Practitioner Spending Grows to Nearly 6 Billion a Year." *Yoga Journal* 26 February 2008.

Maguire, Jennifer Smith. *Fit for Consumption: Sociology and the Business of Fitness.* New York: Routledge, 2008. The quote "Fitness in contemporary society is primarily a commercial enterprise" is paraphrased from this book.

McCullough, Michael. "Lululemon: The Karma Offensive." City News Toronto 7 June 2010. www.citytv.com/toronto/citynews/life/money/article/78614--lululemon-the-karma-offensive.

McGinn, Dave. "How Accurate Are Calorie Counters?" *Globe and Mail* 25 October 2010. The inaccuracies of calorie counters on fitness machines.

Meroni, Roberto, et al. "Comparison of Active Stretching Technique and Static Stretching Technique on Hamstring Flexibility." *Clinical Journal of Sport Medicine* 20.1 (2010): 8–14.

Metcalf, Brad S., et al. "Fatness Leads to Inactivity, But Inactivity Does Not Lead to Fatness: A Longitudinal Study in Children (EarlyBird 45)." *Archives of Disease in Childhood* (2010): no page. First published online.http://adc.bmj.com/content/early/2010/06/23/adc.2009.175927.abstract.

Nesser, Thomas W., et al. "The Relationship between Core Stability and Performance in Division I Football Players." *Journal of Strength and Conditioning Research* 22 (2008): 1750–54.

O'Donovan, Gary, and Rob Shave. "British Adults' Views on the Health Benefits of Moderate and Rigorous Activity." *Preventive Medicine* 45 (2007): 432–35.

O'Mahony, Barry, and John Hall. "The Influence of Perceived Body Image, Vanity and Personal Values on Food Consumption and Related Behaviour." *Journal of Hospitality and Tourism Management* 14 (2007): 57–69.

Peterson, Mark, and Paul Gordon. "Resistance Exercise for the Aging Adult: Clinical Implications and Prescription Guidelines." *American Journal of Medicine* 124.3 (2011): 194–98.

"Physical Activity: Enjoy an Active, Healthy Lifestyle through Physical Activity and Nutrition." Coca-Cola Company (2010). www.thecoca-colacompany.com/citizenship/move.html.

Powell, Kenneth E., Amanda E. Paluch, and Steven N. Blair. "Physical Activity for Health: What Kind? How Much? How Intense? On Top of What?" *Annual Review of Public Health* 32 (2011): 349–65. Nice review of what the evidence says about the value of exercise, including the health benefits of going from sedentary to light activity.

Preidt, R. "Resistance Exercise May Offer Different Cardio Benefits. Weight Training Increased Blood Flow to Limbs More Than Aerobic Exercise Did, Study Finds." MSN.com 10 November 2010. http://health.msn.com/health-topics/alzheimers-disease/articlepage.aspx?cp-documentid=100266924. News coverage of study about benefits of resistance training over aerobic.

Public Health Agency of Canada. www.phac-aspc.gc.ca.

Reynolds, Gretchen. "Weighing the Evidence on Exercise." *New York Times Magazine* 16 April 2010. Review of emerging evidence on benefits of exercise in the context of weight loss and quote from Ravussin.

———. "Phys Ed: The Benefits of Weight Training for Children." *New York Times* 24 November 2010. http://well.blogs.nytimes.com/2010/11/24/phys-ed-the-benefits-of-weight-training-for-kids/?pagemode=print. Review of benefits of weight training for children.

Rodgers, W.M., et al. "Becoming a Regular Exerciser: Examining Change in Behavioural Regulations among Exercise Initiates." *Psychology of Sport and Exercise* 11 (2010): 378–86.

Sacks, Danielle. "Lululemon's Cult of Selling." *Fast Company* 1 April 2009. www.fastcompany.com/magazine/134/om-my.html.

Schaefer, Paul, and Julie Brennan. "Psychological Benefits of Exercise." *Family Physician* 61 (2009): 30–31.

Schjerve, Inga E., et al. "Both Aerobic Endurance and Strength Training Programmes Improve Cardiovascular Health in Obese Adults." *Clinical Science* 115 (2008): 283–93.

Segar, Michelle L., et al. "Midlife Women's Physical Activity Goals: Sociocultural Influences and Effects on Behavioral Regulation." *Sex Roles* 57 (2007): 837–49.

Shenoy, Shweta, Ekta Arora, and Sandhu Jaspal. "Effects of Progressive Resistance Training and Aerobic Exercise on Type 2 Diabetics in Indian Population." *International Journal of Diabetes and Metabolism* 17 (2009): 27–30.

Stewart, Monte. "Lululemon Boss Rides Creative Wave: Athletic Clothier Keeping Abreast of Social Trends and Changes." *Business Edge* 8.10 (2008). www.businessedge.ca/archives/article.cfm/lululemon-boss-rides-creative-wave-17862.

Strelan, Peter, Sarah J. Mehaffey, and Marika Tiggemann. "Self-Objectification and Esteem in Young Women: The Mediating Role of Reasons for Exercise" *Sex Roles* 48 (2003): 89–95.

Swedish National Institute of Public Health. *Physical Activity in the Prevention and Treatment of Disease* (2010). Wonderful summary of the benefits of physical activity that includes a short discussion of how many calories muscle burns ("1 kg muscle has a basal metabolic rate of 10 kcal per day"). This report also notes that strength training "can be a significant aid for the control of body weight" (though the mechanisms are unclear) and is not associated with an increase in muscle mass alone.

Thacker, Stephen B., et al. "The Impact of Stretching on Sports Injury Risk: A Systematic Review of the Literature." *Medicine and Science in Sports and Exercise* 36.3 (2004): 371–78.

Tjønna, Arnt Erik. "Aerobic Interval Training versus Continuous Moderate Exercise as a Treatment for the Metabolic Syndrome: A Pilot Study." *Circulation* 118 (2008): 346–54.

Urstadt, Bryan. "Lust for Lulu: How the Yoga Brand Lululemon Turned Fitness into a Spectator Sport." *New York Times Magazine* 26 July 2009.

Vaucher, Andréa R. "Lululemon Athletica's Yoga-Inspired Sports Attire Positions Itself for Success." *New York Times* 9 September 2007.

Wahba, Phil. "What Are You Doing Today? Not Exercising." MSNBC.com (16 September 2010). http://msnbc-fitness.polls.newsvine.com/_question/2010/09/16/5122996-did-you-exercise-today.

Wahl, Michael J., and David G. Behm. "Not All Instability Training

Devices Enhance Muscle Activation in Highly Resistance-Trained Individuals." *Journal of Strength and Conditioning Research* 22 (2008): 1360–70.

Warburton, Darren, Crystal Whitney Nicol, and Shannon Bredin. "Health Benefits of Physical Activity: The Evidence." *Canadian Medical Association Journal* 174 (2006): 801–9.

Warburton, Darren, et al. "A Systematic Review of the Evidence for Canada's Physical Activity Guidelines for Adults." *International Journal of Behavioral Nutrition and Physical Activity* 7 (2010): 39. Review of the health benefits of exercise.

Weiler, Richard, Emmanel Stamatakis, and Steven Blair. "Should Health Policy Focus on Physical Activity Rather Than Obesity? Yes." *British Medical Journal* 340 (2010): c2603.

Williams, Paul T. "Evidence for the Incompatibility of Age-Neutral Overweight and Age-Neutral Physical Activity Standards from Runners." *American Journal of Clinical Nutrition* 65 (1997): 1391–96.

Winchester, Jason B., et al. "Static Stretching Impairs Sprint Performance in Collegiate Track and Field Athletes." *Journal of Strength and Conditioning Research* 22.1 (2008): 13–19.

Wisløff, Ulrik, et al. "Superior Cardiovascular Effect of Aerobic Interval Training versus Moderate Continuous Training in Heart Failure Patients: A Randomized Study." *Circulation* 115 (2007): 3086–94.

Yamamoto, Linda M., et al. "The Effects of Resistance Training on Endurance Distance Running Performance among Highly Trained Runners: A Systematic Review." *Journal of Strength and Conditioning Research* 22.6 (2008): 2036–44.

———. "The Effects of Resistance Training on Road Cycling Performance among Highly Trained Cyclists: A Systematic Review." *Journal of Strength and Conditioning Research* 24.2 (2010): 560–66.

CHAPTER 2: DIET

Abraham, Carolyn. "Skinny Genes: How DNA Shapes Weight-Loss Success." *Globe and Mail* 17 January 2011.

Adams, Stephen. "Skipping Sleep Halves Fat Loss: Dieters Who Want to Lose Fat Need to Get a Full Night's Sleep, a Study Has Found." *Telegraph* 5 October 2010.

Alexander, Harriet. "Ice Cream, Bacon: New Superfoods." *Edmonton Journal* 24 November 2009.

Alston, Julian M., and Abigail Okrent. "Farm Commodity Policy and Obesity." *Diet and Obesity: Role of Prices and Policies*. International Association of Agricultural Economists, 2009 Pre-Conference Workshop, 16 August 2009. http://ageconsearch.umn.edu/bitstream/53336/2/Alston%20et%20el.pdf.

Alston, Julian M., Daniel A. Sumner, and Stephen A. Vosti. "Farm Subsidies and Obesity in the United States." *Agriculture and Resource Economics Update* 11.2 (2007): 1–4.

Baker, Elizabeth A., et al. "The Role of Race and Poverty in Access to Foods That Enable Individuals to Adhere to Dietary Guidelines." *Preventing Chronic Disease* 3.3 (2006). www.cdc.gov/pcd/issues/2006/jul/05_0217.htm.

Berg, Christina, et al. "Eating Patterns and Portion Size Associated with Obesity in a Swedish Population." *Appetite* 52 (2009): 21–26.

Berman, Mark, and Risa Lavizzo-Mourey. "Obesity Prevention in the Information Age: Caloric Information at the Point of Purchase." *Journal of the American Medical Association* 300.4 (2008): 433–35. Calorie consumption outside the home and underestimation of calories consumed at restaurants.

Birch, Leann L., Jennifer Orlet Fisher, and Kirsten Krahnstoever Davison. "Learning to Overeat: Maternal Use of Restrictive Feeding Practices Promotes Girls' Eating in the Absence of Hunger." *Clinical Nutrition* 78 (2003): 215–20.

Bleich, Sara N., et al. "Increasing Consumption of Sugar-Sweetened Beverages among U.S. Adults: 1988–1994 to 1999–2004." *American Journal of Clinical Nutrition* 89 (2009): 372–81.

Bleich, Sara N., and Keshia M. Pollack. "The Publics' Understanding of Daily Caloric Recommendations and Their Perceptions of Calorie Posting in Chain Restaurants." *BMC Public Health* 10 (2010): 121.

Block, Jason P., Richard A. Scribner, and Karen B. De Salvo. "Fast Food,

Race/Ethnicity, and Income: A Geographic Analysis." *American Journal of Preventive Medicine* 27.3 (2004): 211–17.

Butryn, Meghan L., et al. "Consistent Self Monitoring of Weight: A Key Component of Successful Weight Loss Maintenance." *Obesity* 15 (2007): 3091–96.

"Canada's Food Guide: A Nutritional Juggling Act." CBC News 5 February 2007. www.cbc.ca/news/background/foodguide.

Cannon, Geoffrey. "Why the Bush Administration and the Global Sugar Industry Are Determined to Demolish the 2004 WHO Global Strategy on Diet, Physical Activity and Health." *Public Health Nutrition* 7 (2004): 369–80. Regarding the role of "Big Sugar."

Chandon, Pierre, and Brian Wansink. "The Biasing Health Halos of Fast-Food Restaurant Health Claims: Lower Calorie Estimates and Higher Side-Dish Consumption Intentions." *Journal of Consumer Research* 34 (2007): 301–14.

Chen, Liwei, et al. "Reduction in Consumption of Sugar-Sweetened Beverages Is Associated with Weight Loss: The PREMIER Trial." *American Journal of Clinical Nutrition* 89 (2009): 1299–1306.

Chockalingam, Arun. "Impact of World Hypertension Day." *Canadian Journal of Cardiology* 23.7 (2007): 517–19.

Cohen, Deborah A., and Thomas A. Farley. "Eating as an Automatic Behavior." *Preventing Chronic Disease* 5.1 (2008): 1–7.

Cullen, Karen Weber, et al. "Intake of Soft Drinks, Fruit-Flavored Beverages, and Fruits and Vegetables by Children in Grades 4 through 6." *American Journal of Public Health* 92.9 (2002): 1475–77.

Culp, Jennifer, Robert A. Bell, and Diana Cassady. "Characteristics of Food Industry Web Sites and 'Advergames' Targeting Children." *Journal of Nutrition Education and Behavior* 42.3 (2010): 197–201.

Dansinger, Michael L., et al. "Comparison of the Atkins, Ornish, Weight Watchers, and Zone Diets for Weight Loss and Heart Disease Risk Reduction: A Randomized Trial." *Journal of the American Medical Association* 293 (2005): 43–53.

Devlin, Kate. "Healthy Snacks 'Contain More Sugar Than Ice Cream.'" *Telegraph* 10 May 2010.

"Diet and Physical Activity: A Public Health Priority." World Health Organization (2010). www.who.int/dietphysicalactivity/en/index.html.

Dotinga, Randy. "Low Carb-, Low-Fat Diets Tied for Long-Term Weight Loss." *Businessweek* 2 August 2010. www.businessweek.com/lifestyle/content/healthday/641741.html.

Drewnowski, Adam, and Nicole Darmon. "Food Choices and Diet Costs: An Economic Analysis." *Journal of Nutrition* 135 (2005): 900–4.

Druckman, Ed. "George Bush's New Iraq Plan Includes McDonald's of Baghdad." Associated Content (22 January 2007). www.associatedcontent.com/article/123031/george_bushs_new_iraq_plan_includes.html?cat=7.

Edelson, Ed. "Moderate Coffee, Tea Drinking Lowers Heart Disease Risk." *MSN Health and Fitness* 2 March 2010.

"Families Are 'Lovin' It': Parents' Work Influences How Often Family Meals Are Eaten Outside of Home." *ScienceDaily* 7 May 2011. www.sciencedaily.com/releases/2011/05/110506073825.htm. Amount of family budget spent in restaurants.

Fisher, Jennifer Orlet, and Leann Lipps Birch. "Restricting Access to Palatable Foods Affects Children's Behavioral Response, Food Selection, and Intake." *American Journal of Clinical Nutrition* 69 (1999): 1264–72.

Fowler, Sharon P. "Fueling the Obesity Epidemic? Artificially Sweetened Beverage Use and Long-Term Weight Gain." *Obesity* 16 (2008): 1894–1900. Study showing correlation between diet pop/soda and weight gain.

Francesco, Sofi, et al. "Adherence to Mediterranean Diet and Health Status: Meta-Analysis." *BMJ* 337 (2008): a1344.

Frassetto, Lynda A., et al. "Metabolic and Physiologic Improvements from Consuming a Paleolithic, Hunter-Gatherer Type Diet." *European Journal of Clinical Nutrition* 63 (2009): 947–55. Example study regarding the benefits of a Paleolithic diet.

Gal, David, and Wendy Liu. "Grapes of Wrath: The Angry Effects of Self Control." *Journal of Consumer Research* (2011) (forthcoming). Dieting makes us angry.

Gallagher, Dympna, et al. "Healthy Percentage Body Fat Ranges: An Approach for Developing Guidelines Based on Body Mass Index." *American Journal of Clinical Nutrition* 72 (2000): 694–701. Notes that 11–23 percent for individuals 40–55 is healthy and that anything below that might be considered underweight.

Gorber, S. Connor, et al. "A Comparison of Direct vs. Self-Report Measures for Assessing Height, Weight and Body Mass Index: A Systematic Review." *Obesity Reviews* 8.4 (2007): 307–26.

Gray, Nathan. "Sugar-Sweetened Drink's Diabetes Link 'Clear and Consistent': Meta-Analysis." Food Navigator 29 October 2010. www.foodnavigator-usa.com/Science-Nutrition/Sugar-sweetened-drinks-diabetes-link-clear-and-consistent-Meta-analysis.

Grier, Sonya A., et al. "Fast-Food Marketing and Children's Fast-Food Consumption: Exploring Parents' Influences in an Ethnically Diverse Sample." *Journal of Public Policy and Marketing* 26.2 (2007): 221–35.

Guallar, Eliseo, et al. "Vitamin D Supplementation in the Age of Lost Innocence." *Annals of Internal Medicine* 152 (2010): 327–29. Supports the claim that mortality and morbidity increase with the use of some supplements.

Harris, Jennifer L., et al. "A Crisis in the Marketplace: How Food Marketing Contributes to Childhood Obesity and What Can Be Done." *Annual Review of Public Health* 30 (2009): 211–25.

"Healthy Youth: Nutrition." National Center for Chronic Disease Prevention and Health Promotion (2008). www.cdc.gov/healthyyouth/nutrition/index.htm.

Hebert, James, et al. "Differences between Estimated Caloric Requirements and Self-Reported Caloric Intake in the Women's Health Initiative." *Annals of Epidemiology* 13.9 (2003): 629–37.

Hensley, Wayne. "The Measurement of Height." *Adolescence* 33.131 (1998): 629–35.

Hernandez, Teri L., et al. "Fat Redistribution Following Suction Lipectomy: Defense of Body Fat and Patterns of Restoration." *Obesity* 19 (2011): 1388–95. Study on effects of liposuction.

Hess-Pace, Kate M. "Junk (Food) Policy: The Failure of Nutrition Policy under the United States Department of Agriculture." *The Current* 12.2 (2009): 93–108. "These battles occur, in part, because nutritional education and guidelines cannot contradict the USDA's dominant goal of supporting the business of agriculture" (100).

Institute of Medicine of the National Academies, Committee on Food Marketing and the Diets of Children and Youth. *Food Marketing to Kids: Threat or Opportunity.* Edited by J. Michael McGinnis, Jennifer Appleton Gootman, and Vivica I. Kraak. Washington, DC: The National Academies Press, 2006.

James, Larry. "Why Do Some People Lose Weight and Keep It Off? Ten Common Steps for Successful Weight Loss over the Lifespan." In *Handbook of Obesity Intervention for the Lifespan*, ed. Larry C. James and John C. Linton, 85–92. New York: Springer, 2009. Factors that contribute to long-term weight maintenance.

Johansson, Gunnar, et al. "Underreporting of Energy Intake in Repeated 24-Hour Recalls Related to Gender, Age, Weight Status, Day of Interview, Educational Level, Reported Food Intake, Smoking Habits and Area of Living." *Public Health Nutrition* 4 (2001): 919–27.

Johnson, Paul M., and Paul J. Kenny. "Dopamine D2 Receptors in Addiction-Like Reward Dysfunction and Compulsive Eating in Obese Rats." *Nature Neuroscience* 13 (2010): 635–41.

Katan, Martin. "Weight-Loss Diets for the Prevention and Treatment of Obesity." *New England Journal of Medicine* 360 (2009): 923–24.

Keast, Debra R., Theresa A. Nicklas, and Carol E. O'Neil. "Snacking Is Associated with Reduced Risk of Overweight and Reduced Abdominal Obesity in Adolescents: National Health and Nutrition Examination Survey (NHANES) 1999–2004." *American Journal of Clinical Nutrition* 92 (2010): 428–35.

Kimmons, Joel, et al. "Fruit and Vegetable Intake among Adolescents and Adults in the United States: Percentage Meeting Individualized Recommendations." *Medscape Journal of Medicine* 11.1 (2009): 26.

Kondro, Wayne. "Proposed Canada Food Guide Called 'Obesogenic.'" *Canadian Medical Association Journal* 174.5 (2006): 605–6.

Krebs-Smith, Susan M., et al. "Americans Do Not Meet Federal Dietary Recommendations." *Journal of Nutrition* 140 (2010): 1832–38.

Lee, I-Min, et al. "Vitamin E in the Primary Prevention of Cardiovascular Disease and Cancer. The Women's Health Study: A Randomized Controlled Trial." *Journal of the American Medical Association* 294 (2005): 56–65.

Lesser, Lenard I., et al. "Relationship between Funding Source and Conclusion among Nutrition-Related Scientific Articles." *PLoS Medicine* 4.1 (2007): 41–46.

Leung, Wency. "U.S. Burger Joint Unleashes 1,500-Calorie Sandwich." *Globe and Mail* 22 June 2010.

Lichtman, Steven, et al. "Discrepancy between Self-Reported and Actual Caloric Intake and Exercise in Obese Subjects." *New England Journal of Medicine* 327.27 (1992): 1893–98.

Lippman, Scott M., et al. "Effect of Selenium and Vitamin E on Risk of Prostate Cancer and Other Cancers. The Selenium and Vitamin E Cancer Prevention Trial (SELECT)." *Journal of the American Medical Association* 301.1 (2009): 39–51.

Lubow, Arthur. "Steal This Burger." *New York Times* 19 April 1998.

Ludwig, David, and Marion Nestle. "Can the Food Industry Play a Constructive Role in the Obesity Epidemic?" *Journal of the American Medical Association* 300 (2008): 1808–10.

Lund University. "Coffee May Protect against Breast Cancer, Study Shows." *ScienceDaily* 25 April 2008. www.sciencedaily.com/releases/2008/04/080424115324.htm.

MacLean, Catherine H., et al. "Effects of Omega-3 Fatty Acids on Cancer Risk: A Systematic Review." *Journal of the American Medical Association* 295 (2006): 403–15.

Malik, Vasanti S., Matthias B. Schulze, and Frank B. Hu. "Intake of Sugar-Sweetened Beverages and Weight Gain: A Systematic Review." *American Journal of Clinical Nutrition* 84 (2006): 274–88.

Mikolajczyk, Rafael T., et al. "Relationship between Perceived Body Weight and Body Mass Index Based on Self-Reported Height and Weight among University Students: A Cross-Sectional

Study in Seven European Countries." *BMC Public Health* 10 (2010): 40.

Mulholland, Angela. "Like Heroin, Junk Food May 'Hijack' Our Brains." CTV Toronto 26 May 2010. www.ctv.ca/CTVNews/Health/20100514/psychology-of-food-3-100526.

National Weight Control Registry (NWCR). Brown Medical School/The Miriam Hospital, Weight Control and Diabetes Research Center. www.nwcr.ws.

Nestle, Marion. "Food Lobbies, the Food Pyramid, and U.S. Nutrition Policy." *International Journal of Health Services* 23.3 (1993): 483–96.

———. "Food Company Sponsorship of Nutrition Research and Professional Activities: A Conflict of Interest?" *Public Health Nutrition* 4 (2001): 1015–22.

———. *Food Politics: How the Food Industry Influences Nutrition and Health*. Berkeley: University of California Press, 2007.

Nettleton, Jennifer, et al. "Diet Soda Intake and Risk of Incident Metabolic Syndrome and Type 2 Diabetes in the Multi-Ethnic Study of Atherosclerosis (MESA)." *Diabetes Care* 32 (2009): 688–94. Example study of the association between the consumption of soda/pop and health issues.

"New U.S. Dietary Guidelines: Progress, Not Perfection." Harvard School of Public Health (2011). www.hsph.harvard.edu/nutrition source/what-should-you-eat/dietary-guidelines-2010/index.html.

Phelan, Suzanne, et al. "Recovery from Relapse among Successful Weight Maintainers." *American Journal of Clinical Nutrition* 78.6 (2003): 1079–84.

Pillitteri, Janine L., et al. "Use of Dietary Supplements for Weight Loss in the United States: Results of a National Survey." *Journal of the American Medical Association* 249 (2005): 56–65.

"Politics Too Influential in New Dietary Guidelines, Says Nutrition Expert." Food Navigator (2011). www.foodnavigator.com. Nestle quote about industry not liking the "eat less" message.

Provencher, Véronique, Janet Polivya, and C. Peter Hermana. "Perceived Healthiness of Food. If It's Healthy, You Can Eat More!" *Appetite* 52.2 (2009): 340–44.

Purslow, Lisa, et al. (including N. Wareham). "Energy Intake at Breakfast and Weight Change: Prospective Study of 6,764 Middle-aged Men and Women." *American Journal of Epidemiology* 167 (2008): 188–92.

Rabin, Roni Caryn. "Nutrition: Study Examines a Diet from TV Ads." *New York Times* 4 June 2010.

Reedy, Jill, and Susan Krebs-Smith. "A Comparison of Food-Based Recommendations and Nutrient Values of Three Food Guides: USDA's MyPyramid, NHLBI's Dietary Approaches to Stop Hypertension Eating Plan, and Harvard's Healthy Eating Pyramid." *Journal of the American Dietetic Association* 108 (2008): 522–28.

"Report of the Dietary Guidelines Advisory Committee on the Dietary Guidelines for Americans." United States Department of Agriculture (2010). www.cnpp.usda.gov/DGAs2010-DGACReport.htm.

Rettner, Rachael. "Organic Labels May Trick Dieters into Overeating." *Live Science* 24 June 2010.

Rhee, Kyung E., et al. "Parenting Styles and Overweight Status in First Grade." *Pediatrics* 117.6 (2006): 2047–54.

Roberto, Christina A., et al. "Evaluating the Impact of Menu Labeling on Food Choices and Intake." *American Journal of Public Health* 100 (2010): 312–18.

Robinson, Thomas N., and Donna M. Matheson. "Effects of Fast Food Branding on Young Children's Taste Preferences." *Archives of Pediatrics and Adolescent Medicine* 161.8 (2007): 792–97.

Rolls, Barbara J., Liane S. Roe, and Jennifer S. Meengs. "Salad and Satiety: Energy Density and Portion Size of a First-Course Salad Affect Energy Intake at Lunch." *Journal of the American Dietetic Association* 104.10 (2004): 1570–76.

Romaguera, Dora, et al. (including N. Wareham). "Adherence to Mediterranean Diet Is Associated with Lower Abdominal Adiposity in European Men and Women." *Journal of Nutrition* 139 (2009): 1728–37.

Rosenheck, Robert. "Fast Food Consumption and Increased Caloric Intake: A Systematic Review of a Trajectory Towards Weight Gain and Obesity Risk." *Obesity Reviews* 9.6 (2008): 535–47.

Rozin, Paul, et al. "The Ecology of Eating: Smaller Portion Sizes in France Than in the United States Help Explain the French Paradox." *Psychological Science* 14.5 (2003): 450–54.

Sacks, Frank M., et al. "Comparison of Weight-Loss Diets with Different Compositions of Fat, Protein and Carbohydrates." *New England Journal of Medicine* 360 (2009): 859–73.

Scagliusi, Fernanda B., et al. "Selective Underreporting of Energy Intake in Women: Magnitude, Determinants, and Effect of Training." *Journal of the American Dietetic Association* 103.10 (2003): 1306–13.

Schwartz, Marlene B., et al. "Examining the Nutritional Quality of Breakfast Cereals Marketed to Children." *Journal of the American Dietetic Association* 108 (2008): 702–5.

Shai, Iris, et al. "Weight Loss with a Low-Carbohydrate, Mediterranean, or Low-Fat Diet." *New England Journal of Medicine* 359 (2008): 229–41.

Shields, Margot, Sarah Connor Gorber, and Mark S. Tremblay. "Methodological Issues in Anthropometry: Self-Reported versus Measured Height and Weight." Statistics Canada (2009). www.statcan.gc.ca/pub/11-522-x/2008000/article/11002-eng.pdf.

Shields, M., M.D. Carroll, and C.L. Ogden. "Adult Obesity Prevalence in Canada and the United States." NCHS Data Brief, no. 56. Hyattsville, MD: National Center for Health Statistics, 2011. Study released jointly by the US Centers for Disease Control and Statistics Canada that found that 24.1 percent of adults in Canada and 34.4 percent of adult Americans were obese between 2007 and 2009.

Shields, Margot, Sarah Connor Gorber, and Mark S. Tremblay. "Estimates of Obesity Based on Self-Report versus Direct Measures." *Health Reports* 19.2 (June 2008). Statistics Canada study on self-reported height and weight.

Slatore, Christopher G., et al. "Long-Term Use of Supplemental Multivitamins, Vitamin C, Vitamin E, and Folate Does Not Reduce the Risk of Lung Cancer." *American Journal of Respiratory and Critical Care Medicine* 177 (2008): 524–30.

Smoyer-Tomica, Karen E., et al. "The Association between Neighborhood Socioeconomic Status and Exposure to Supermarkets and Fast Food Outlets." *Health and Place* 14.4 (2008): 740–54.

Speers, Sarah, et al. *Public Perceptions of Food Marketing to Youth: Results of the Rudd Center Public Opinion Poll, May 2008*. Rudd Center for Food Policy and Obesity, Yale University, 2009.

Steenhuis, Ingrid H.M., and Willemijn M. Vermeer. "Portion Size: Review and Framework for Interventions." *International Journal of Behavioral Nutrition and Physical Activity* 6 (2009): 58.

Steenhuysen, Julie. "Study Suggests Processed Meat a Real Health Risk." *Edmonton Journal* 18 May 2010.

"Survey Identifies Diet Fatigue as a Leading Cause of Diet Failure." *Medical News Today* 18 October 2007. Report on a survey of number of diets tried.

Symons, Jane. "Which Supplements Should YOU Take?" *Mail Online* 28 September 2010. www.dailymail.co.uk/health/article-1315772/Which-supplements-YOU-take.html. Summary of pros and cons of popular supplements.

Tjepkema, Michael. "Measured Obesity: Adult Obesity in Canada: Measured Height and Weight." Statistics Canada (2005). www.statcan.gc.ca/pub/82-620-m/2005001/pdf/4224906-eng.pdf.

Trichopoulou, Antonia, Christina Bamia, and Dimitrios Trichopoulos. "Anatomy of Health Effects of Mediterranean Diet: Greek EPIC Prospective Cohort Study." *BMJ* 338 (2009): b2337.

Tsai, Adam Gilden, and Thomas A. Wadden. "Systematic Review: An Evaluation of Major Commercial Weight Loss Programs in the United States." *Annals of Internal Medicine* 142 (2005): 56–66.

Urban L.E., G.E. Dallal, L.M. Robinson, L.M. Ausman, E. Saltzman, and S.B. Roberts. "The Accuracy of Stated Energy Contents of Reduced-Energy, Commercially Prepared Foods." *Journal of the American Dietetic Association* 110 (2010): 116–23. Calorie labels often inaccurate.

US Department of Agriculture and US Department of Health and Human Services. *Dietary Guidelines for Americans, 2010.* 7th ed. Washington, DC: US Government Printing Office, December 2010.

Wang, Chenchen, et al. "n-3 Fatty Acids from Fish or Fish-Oil Supplements, but not -Linolenic Acid, Benefit Cardiovascular Disease Outcomes in Primary- and Secondary-Prevention Studies: A Systematic Review." *American Journal of Clinical Nutrition* 84 (2006): 5–17.

Wang, Lu, et al. "Systematic Review: Vitamin D and Calcium Supplementation in Prevention of Cardiovascular Events." *Annals of Internal Medicine* 152 (2010): 315–23.

Wang, Y. Claire, Sara N. Bleich, and Steven L. Gortmaker. "Increasing Caloric Contribution from Sugar-Sweetened Beverages and 100 Percent Fruit Juices Among U.S. Children and Adolescents, 1988–2004." *Pediatrics* 121 (2008): e1604–e1614.

Wansink, Brian, and Junyong Kim. "Bad Popcorn in Big Buckets: Portion Size Can Influence Intake as Much as Taste." *Journal of Nutrition Education and Behavior* 37 (2005): 242–45.

Wansink, Brian, James E. Painter, and Jill North. "Bottomless Bowls: Why Visual Cues of Portion Size May Influence Intake." *Obesity Research* 13 (2005): 93–100.

Weeks, Carly. "Granola Bars: A Healthy Snack or Dressed-Up Junk Food?" *Globe and Mail* 16 March 2011. Quote on granola bars "masquerading as health food."

Wilkinson, Susan B.T., Gene Rowe, and Nigel Lambert. "The Risks of Eating and Drinking: Consumer Perceptions and 'Reality.'" *EMBO Reports* 5.S1 (2004): S27–S31.

Willett, Walter. "The Mediterranean Diet: Science and Practice." *Public Health Nutrition* 9 (2006): 105–10.

Wisdom, Jessica, Julie S. Downs, and George Loewenstein. "Promoting Healthy Choices: Information versus Convenience." *American Economic Journal: Applied Economics* 2.2 (2010): 164–78.

World Health Organization. *Obesity and Overweight.* Fact Sheet no. 311 (updated March 2011).

Wyatt, Holly R., et al. "Long-Term Weight Loss and Breakfast in Subjects in the National Weight Control Registry." *Obesity Research* 10 (2002): 78–82.

Yon, Bethany A., et al. "Personal Digital Assistants Are Comparable to Traditional Diaries for Dietary Self-Monitoring during a Weight Loss Program." *Journal of Behavioral Medicine* 30.2 (2007): 165–75.

Young, Lisa R., and Marion Nestle. "Portion Sizes and Obesity: Responses of Fast-Food Companies." *Journal of Public Health Policy* 28 (2007): 238–48. Research on the portion size of fast food restaurants.

"You're Out, You're Hungry, You're Doin' Fourthmeal." Taco Bell. www.tacobell.com/fourthmeal.

CHAPTER 3: GENETICS

Bubela, Tania M., and Timothy Caulfield. "Do the Print Media 'Hype' Genetic Research? A Comparison of Newspaper Stories and Peer-Reviewed Research Papers." *Canadian Medical Association Journal* 170.9 (2004): 1399–1407.

———. "Role and Reality: Technology Transfer at Canadian Universities." *Trends in Biotechnology* 28.9 (2010): 447–51.

Campion, Edward W. "Medical Research and the News Media." *New England Journal of Medicine* 351 (2004): 2436–37. How media report medical research.

Canada. Canadian Institutes of Health Research Act, R.S.C. 2000, c.6.

Caulfield, Timothy. "Biotechnology and the Popular Press: Hype and the Selling of Science." *Trends in Biotechnology* 22.7 (2004): 337–39.

———. "Popular Media, Biotechnology, and the 'Cycle of Hype.'" *Houston Journal of Health Law and Policy* 5 (2005): 213–33.

Caulfield, Timothy, et al. "Direct-to-Consumer Genetic Testing: Good, Bad or Benign?" *Clinical Genetics* 77 (2010): 101–5. Challenges associated with direct-to-consumer testing.

Collins, Francis S. "Medical and Societal Consequences of the Human Genome Project." *New England Journal of Medicine* 341 (1999): 28–37.

———. "Has the Revolution Arrived?" *Nature* 464 (2010): 674–75.

Collins, Francis S., and Monique K. Mansoura. "The Human Genome

Project: Revealing the Shared Inheritance of All Humankind Cancer." *Cancer* 91 (2001): 221–25. U.S. Human Genome Project Research Goals. www.ornl.gov/sci/techresources/Human_Genome/hg5yp/index.shtml.

de Semir, Vladimir, Cristina Ribas, and Gemma Revuelta. "Press Releases of Science Journal Articles and Subsequent Newspaper Stories on the Same Topic." *Journal of the American Medical Association* 280 (1998): 294–95. The impact of press releases on how the media covers science.

Drmanac, Radoje. "Human Genome Sequencing Using Unchained Base Reads on Self-Assembling DNA Nanoarrays." *Science* 327 (2010): 78–79.

Fujimura, Joan H. "The Molecular Biological Bandwagon in Cancer Research: Where Social Worlds Meet." *Social Problems* 35.3 (1988): 261–83.

Gillis, Charlie. "Doomed Newfoundlanders Opt to Eat, Drink and Be Merry." *National Post* 12 April 1999. Story on predisposition to heart disease and the population's "fatalistic" response.

Gray, Richard. "World Still Waiting for Genome DNA Map to Unlock Secrets." *Daily Telegraph* 20 June 2010.

Gundle, Kenneth R., Molly J. Dingel, and Barbara A. Koenig. "'To Prove This Is the Industry's Best Hope': Big Tobacco's Support of Research on the Genetics of Nicotine Addiction." *Addiction* 105 (2010): 974–83.

Hall, Stephen S. "Revolution Postponed: Why the Human Genome Project Has Been Disappointing." *Scientific American* 303.4 (2010): 60–67.

Hall, Wayne. "British Drinking: A Suitable Case for Treatment?" *BMJ* 331 (2005): 527–28.

Hall, Wayne D., Rebecca Mathews, and Katherine I. Morley. "Being More Realistic about the Public Health Impact of Genomic Medicine." *PLoS Medicine* 7.10 (2010). www.plosmedicine.org/article/info%3Adoi%2F10.1371%2Fjournal.pmed.1000347. The lack of value of genomics to public health and the possible impact of the tobacco, alcohol, and gambling industries on the profile of genetics.

Heshka, Jodi T., et al. "A Systematic Review of Perceived Risks, Psychological and Behavioral Impacts of Genetic Testing." *Genetics in Medicine* 10 (2008): 19–32. Analysis of the minimal impact of genetic risk information on behaviour.

Highfield, Roger. "Selling Science to the Public." *Science* 289 (2000): 59.

Janssens, A., and J.W. Cecile, et al. "A Critical Appraisal of the Scientific Basis of Commercial Genomic Profiles Used to Assess Health Risks and Personalized Health Interventions." *American Journal of Human Genetics* 82 (2008): 593–99.

Kimmelman, Jonathan. *Gene Transfer and the Ethics of First-in-Human Research: Lost in Translation*. Cambridge: Cambridge University Press, 2009. History of gene therapy.

Lango, Hana, et al. "Assessing the Combined Impact of 18 Common Genetic Variants of Modest Effect Sizes on Type 2 Diabetes Risk." *Diabetes* 57 (2008): 3129–35. Study on the predictive value of genetic variants.

Lemmens, Trudo, and Duff Waring. *Law and Ethics in Biomedical Research*. Toronto: University of Toronto Press, 2006. Review of the Jesse Gelsinger case.

Lyssenko, Valeriya, et al. "Clinical Risk Factors, DNA Variants, and the Development of Type 2 Diabetes." *New England Journal of Medicine* 359 (2008): 2220–32. Study that demonstrates the predictive limits of genomic risk information.

Magnus, David, Mildred Cho, and Robert Cook-Deegan. "Direct-to-Consumer Genetic Tests: Beyond Medical Regulation?" *Genome Medicine* 1.2 (2009): 17.1–17.3. Discussion of the predictive value of genetic tests.

Marteau, Theresa M., et al. "Effects of Communicating DNA-Based Disease Risk Estimates on Risk-Reducing Behaviours." *Cochrane Database of Systematic Reviews* 10 (2010). www2.cochrane.org/reviews/en/ab007275.html.

Marteau, Theresa M., and Caryn Lerman. "Genetic Risk and Behavioural Change." *BMJ* 322 (2001): 1056–59 at 1057.

Marteau, Theresa M., and John Weinman. "Self-Regulation and the

Behavioural Response to DNA Risk Information: A Theoretical Analysis and Framework for Future Research." *Social Science and Medicine* 62 (2006): 1360–68.

McBride, Colleen M., et al. "The Behavioral Response to Personalized Genetic Information: Will Genetic Risk Profiles Motivate Individuals and Families to Choose More Healthful Behaviors?" *Annual Review of Public Health* 31 (2010): 13.1–13.15.

My DNA Fragrance. http://mydnafragrance.com/perfume.

Ordovas, Jose M., and Vincent Mooser. "Nutrigenomics and Nutrigenetics." *Current Opinion in Lipidology* 15.2 (2004): 101–8.

Pedersen, Nancy L. "Reaching the Limits of Genome-Wide Significance in Alzheimer Disease: Back to the Environment." *Journal of the American Medical Association* 303 (2010): 1864–65. Example of the limited role of genes in the development of complex diseases.

Preidt, Robert. "Most Heart Patients Skimp on Exercise after Rehab: A Year Later, Only 37 Percent Were Doing Cardio Exercises 3 Times a Week, Study Finds." *HealthDay News* 18 June 2010. http://health.usnews.com/health-news/family-health/heart/articles/2010/06/19/most-heart-patients-skimp-on-exercise-after-rehab.

Ransohoff, David F., and Richard M. Ransohoff. "Sensationalism in the Media: When Scientists and Journalists May Be Complicit Collaborators." *Effective Clinical Practice* 4 (2001): 185–88. Use of phrase "complicit collaborators" to describe the relationship between media and research community.

Subcommittee on Human Genome of the Health and Environmental Research Advisory Committee. *Report on the Human Genome Initiative for the Office of Health and Environmental Research.* U.S. Department of Energy, Office of Energy Research, Office of Health and Environmental Research, 27 April 1987. www.ornl.gov/sci/techresources/Human_Genome/project/herac2.shtml.

Wade, Nicholas. "A Dissenting Voice as the Genome Is Sifted to Fight Disease." *New York Times* 15 September 2008. Interview with David Goldstein.

Wallace, Helen M. "Big Tobacco and the Human Genome: Driving the Scientific Bandwagon?" *Genomics, Society and Policy* 5.1 (2009): 80–133.

Wilson, James M. "A History Lesson for Stem Cells." *Science* 324.5928 (2009): 727–28. Expectations leading to premature clinical research.

CHAPTER 4: REMEDIES

Abramson, John, and Barbara Starfield. "The Effect of Conflict of Interest on Biomedical Research and Clinical Practice Guidelines: Can We Trust the Evidence in Evidence-Based Medicine?" *Journal of the American Board of Family Practice* 18 (2005): 414–18.

Associated Press. "$34 Billion Spent Yearly on Alternative Medicine." MSNBC.com 30 July 2009. Amount spent on alternative remedies (11 percent of all out-of-pocket spending).

Association of Accredited Naturopathic Medical Colleges (AANMC) (2010). www.aanmc.org. Regarding accredited schools and curriculum. For interview with Daniel Rubin see www.aanmc.org/careers/alumni-leaders-in-the-field/daniel-rubin-profile.php#philosophy. I am grateful to blogger Orac for pointing out this exchange. http://scienceblogs.com/insolence/2010/09/a_highly_revealing_quote_from_a_naturopa.php.

Astin, John A. "Why Patients Use Alternative Medicine: Results of a National Study." *Journal of the American Medical Association* 279.19 (1998): 1548–53. People use CAM because it fits with their philosophical world view.

Benos, Dale J., et al. "The Ups and Downs of Peer Review." *Advances in Physiology Education* 31 (2007): 145–52.

Berman, Brian M. "The Cochrane Collaboration and Evidence-Based Complementary Medicine." *Journal of Alternative and Complementary Medicine* 3 (1997): 191–94. Definition of CAM. In order to be comparable and inclusive, this paper uses the definition from the Cochrane Collaboration that CAM "includes all such practices and ideas which are outside the domain of conventional medicine in several countries and defined by its users as preventing or treating illness, or promoting health and well-being."

Berman, Brian M., et al. "Acupuncture for Chronic Low Back Pain." *New England Journal of Medicine* 363 (2010): 454–61.

Bhandari, Mohit, et al. "Association between Industry Funding and Statistically Significant Pro-Industry Findings in Medical and Surgical Randomized Trials." *Canadian Medical Association Journal* 170 (2004): 477–80.

Bonevski, Billie, Amanda Wilson, and David A. Henry. "An Analysis of News Media Coverage of Complementary and Alternative Medicine." *PLoS ONE* 3.6 (2008). www.plosone.org/article/info:doi/10.1371/journal.pone.0002406. The Australian study on media coverage of CAM.

Boseley, Sarah. "Ban Homeopathy from NHS, Say Doctors." *Guardian* 29 June 2010.

Brennan, Troyen A., et al. "Health Industry Practices That Create Conflicts of Interest." *Journal of the American Medical Association* 295 (2006): 429–33.

Brett, Allan S., Wayne Burr, and Jamaluddin Moloo. "Are Gifts from Pharmaceutical Companies Ethically Problematic? A Survey of Physicians." *Archives of Internal Medicine* 163 (2003): 2213–18.

Brown, David. "Critics Object to 'Pseudoscience' Center." *Washington Post* 17 March 2009. Review of the push to close NCCAM.

Busse, Jason W., Kumanan Wilson, and James B. Campbell. "Attitudes towards Vaccination among Chiropractic and Naturopathic Students." *Vaccine* 26.49 (2008): 6237–43.

Cancer Cure Foundation. www.cancure.org/homeopathy.htm. An example of a website that is supportive of homeopathy.

Cassileth, Barrie R., and Andrew J. Vickers. "High Prevalence of Complementary and Alternative Medicine Use among Cancer Patients: Implications for Research and Clinical Care." *Journal of Clinical Oncology* 23.12 (2005): 2590–92. Use of CAM among cancer patients.

Caulfield, Timothy. "Profit and the Production of the Knowledge: The Impact of Industry on Representations of Research Results." *Harvard Health Policy Review* 8 (2007): 68–77.

Caulfield, Timothy, and Colin Feasby. "Potions, Promises and Paradoxes: Complementary and Alternative Medicine and Malpractice Law in Canada." *Health Law Journal* 9 (2001): 183–203.

Charlton, Bruce G. "Healing but Not Curing: Alternative Medical Therapies as Valid New Age Spiritual Healing Practices." In *Healing, Hype, or Harm? A Critical Analysis of Complementary or Alternative Medicine,* ed. Edzard Ernst, 68–77. Exeter: Societas, 2008.

Choudry, Niteesh K., Henry T. Stelfox, and Allan S. Detsky. "Relationships between Authors of Clinical Practice Guidelines and the Pharmaceutical Industry." *Journal of the American Medical Association* 287 (2002): 612–17.

Colquhoun, David. "Science Degrees without the Science." *Nature* 446.7134 (2007): 373–74.

———. "Alternative Medicine in UK Universities." In *Healing, Hype, or Harm? A Critical Analysis of Complementary or Alternative Medicine,* ed. Edzard Ernst, 40–67. Exeter: Societas, 2008. Reference to the idea of the "lying dilemma."

"Complaint after NHS Highland's Decision on Homeopathy." BBC News 10 October 2010.

"Complementary Therapies: The Big Con." *Independent* 22 April 2008.

Corinthian Naturopathic College. www.corinthiannaturopathic college.com/course_program.

DeAngelis, Catherine D., Phil B. Fontanarosa, and Annette Flanagin. "Reporting Conflicts of Interest and Relationships between Investigators and Research Sponsors." *Journal of the American Medical Association* 286 (2001): 89–91.

de Bruyn, Theodore. *A Summary of National Data on Complementary and Alternative Health Care—Current Status and Future Development: A Discussion Paper.* Health Canada, Government of Canada, 2002. www.hc-sc.gc.ca/dhp-mps/alt_formats/hpfb-dgpsa/pdf/pubs/cahc-acps-summary-synthesec-eng.pdf. Trends in CAM use in Canada.

Donnelly, Laura. "Homeopathy Is Witchcraft, Say Doctors." *Telegraph* 15 May 2010.

Eggertson, Laura. “Naturopathic Doctors Gaining New Powers.” *Canadian Medical Association Journal* 182 (2010): e29–e30. The status of naturopaths in Canada.

Ernst, Edzard. *The Desktop Guide to Complementary and Alternative Medicine: An Evidence-Based Approach*. 2nd ed. Edinburgh: Mosby, 2006.

———. “Complementary and Alternative Medicine: What the NHS Should Be Funding.” *British Journal of General Practice* 58.548 (2008): 208–9.

———. “Review: St John’s Wort Superior to Placebo and Similar to Antidepressants for Major Depression but with Fewer Side Effects.” *Evidence-Based Mental Health* 12.3 (2009): 78.

Evans, Maggie, et al. “Decisions to Use Complementary and Alternative Medicine (CAM) by Male Cancer Patients: Information-Seeking Roles and Types of Evidence Used.” *BMC Complementary and Alternative Medicine* 7 (2007): 25. Complexity of reasons behind the use of alternative medicine and need for frank disclosure regarding standards of evidence.

Fisher, Morris A. “Physicians and the Pharmaceutical Industry: A Dysfunctional Relationship.” *Perspectives in Biology and Medicine* 4 (2003): 254–72.

Fleming, Sara A., and Nancy C. Gutknecht. “Naturopathy and the Primary Care Practice.” *Primary Care: Clinics in Office Practice* 37 (2010): 119–36. History and principles of naturopathy and the increase in percentage of NDs.

Friedman, Lee S., and Elihu D. Richter. “Relationship between Conflicts of Interest and Research Results.” *Journal of General Internal Medicine* 19 (2004): 51–56.

Fugh-Berman, Adriane J. “The Haunting of Medical Journals: How Ghostwriting Sold ‘HRT.’” *PLoS Medicine* 7.9 (2010). www.plosmedicine.org/article/info%3Adoi%2F10.1371%2Fjournal.pmed.1000335.

Fugh-Berman, Adriane, and Shahram Ahari. “Following the Script: How Drug Reps Make Friends and Influence Doctors.” *PLoS*

Medicine 4.4 (2007). www.plosmedicine.org/article/info:doi/10.1371/journal.pmed.0040150.

Gagnon, Marc-André, and Joel Lexchin. "The Cost of Pushing Pills: A New Estimate of Pharmaceutical Promotion Expenditures in the United States." *PLoS Medicine* 5(1) (2008). www.plosmedicine.org/article/info:doi/10.1371/journal.pmed.0050001.

Goldacre, Ben. "Is the Conflict of Interest Unacceptable When Drug Companies Conduct Trials on Their Own Drugs?" *BMJ* 339 (2009): 1286–87.

Gøtzsche, Peter C., et al. "Ghost Authorship in Industry-Initiated Randomised Trials." *PLoS Medicine* 4.1 (2007): e19, 47–52.

Gunasekera, Varuni, Edzard Ernst, and Daniel George Ezra. "Systematic Internet-Based Review of Complementary and Alternative Medicine for Glaucoma." *Opthalmology* 115.3 (2008): 435–39.

Hall, Harriet. "The Graston Technique—Inducing Microtrauma with Instruments." *Science-Based Medicine* (2010). www.science-basedmedicine.org/?p=3170.

Hammer, Warren I. "The Effect of Mechanical Load on Degenerated Soft Tissue." *Journal of Bodywork and Movement Therapies* 12.3 (2008): 246–56. Study on the GT technique.

Handley, Doug V., Nick A. Rieger, and David J. Rodda. "Rectal Perforation from Colonic Irrigation Administered by Alternative Practitioners." *Medical Journal of Australia* 181.10 (2004): 575–76. Evidence of bowel perforation as a result of colon cleanses.

Hawk, J. Chris. "Influence of Funding Source on Outcome, Validity, and Reliability of Pharmaceutical Research." *Council on Scientific Affairs Report 10*, American Medical Association 2004 Annual Meeting, June 2004. www.ama-assn.org/ama1/pub/upload/mm/443/a04csa10-fulltext.pdf.

House of Commons Science and Technology Committee. "Evidence Check 2: Homeopathy." *UK House of Commons Fourth Report of Session 2009–10*. London: The Stationery Office Limited, 22 February 2010. www.publications.parliament.uk/pa/cm200910/cmselect/cmsctech/45/45.pdf.

Hunt, Katherine, and Edzard Ernst. "Evidence-Based Practice in British Complementary and Alternative Medicine: Double Standards?" *Journal of Health Services Research and Policy* 14 (2009): 219–23.

Industry Funding of Medical Education: Report of an AAMC Task Force. Association of American Medical Colleges, June 2008. http://services.aamc.org/publications/showfile.cfm?file=version114.pdfandprd_id=232.

"International Public Opinion Research on Emerging Technologies: Canada–U.S. Survey Results." Canadian Biotechnology Secretariat, March 2005. http://bioportal.gc.ca/CMFiles/E-POR-ET_200549QZS-5202005-3081.pdf.

Journal of Manipulative and Physiological Therapeutics. www.journals.elsevierhealth.com/periodicals/ymmt/home.

Kemper, Kathi J., et al. "The Use of Complementary and Alternative Medicine in Pediatrics." *Pediatrics* 122 (2008): 1374–86. CAM use in North America.

Kind, Christoph. "Endorsing Naturopathic Medicine Accepts Science over Spin." *Vancouver Sun* 3 March 2009.

Lee, Anna, and Mary Done. "Stimulation of the Wrist Acupuncture Point P6 for Preventing Postoperative Nausea and Vomiting." *Cochrane Database of Systematic Reviews* 3 (2004): CD003281.

Lee, Anna, and Lawrence T.Y. Fan. "Stimulation of the Wrist Acupuncture Point P6 for Preventing Postoperative Nausea and Vomiting." *Cochrane Database of Systematic Reviews* 2 (2009). www.thecochranelibrary.com/userfiles/ccoch/file/.../CD003281.pdf.

Lemmens, Trudo. "Leopards in the Temple: Restoring Scientific Integrity to the Commercialized Research Scene." *Journal of Law, Medicine and Ethics* 32.4 (Winter 2004): 641–57.

Lewis, Tracy R., Jerome H. Reichman, and Anthony D. So. "The Case for Public Funding and Public Oversight of Clinical Trials." *The Economists' Voice* 4.1 (January 2007). http://works.bepress.com/tracy_lewis/2.

Lexchin, Joel, et al. "Pharmaceutical Industry Sponsorship and Research Outcome and Quality: Systematic Review." *BMJ* 326 (2003): 1167.

Lièvre, Michel, et al. "Premature Discontinuation of Clinical Trial for Reasons Not Related to Efficacy, Safety, or Feasibility." *BMJ* 322 (2001): 603–6.

Linde, Klaus, Michael M. Berner, and Levente Kriston. "St John's Wort for Major Depression." *Cochrane Database of Systematic Reviews* 4 (2008): CD000448.

Loghmani, M. Terry, and Stuart J. Warden. "Instrument-Assisted Cross-Fiber Massage Accelerates Knee Ligament Healing." *Journal of Orthopaedic and Sports Physical Therapy* 39.7 (2009): 506–14. Study on GT.

Lourenco, Maria Teresa, et al. "Superstition but Not Distrust in the Medical System Predicts the Use of Complementary and Alternative Medicine in a Group of Patients with Acute Leukemia." *Leukemia and Lymphoma* 49.2 (2008): 339–41.

MacLennan, Alastair H., Stephen P. Myers, and Anne W. Taylor. "The Continuing Use of Complementary and Alternative Medicine in South Australia: Costs and Beliefs in 2004." *Medical Journal of Australia* 184 (2006): 27–31. Use of CAM in Australia and misperceptions about government regulation and testing.

Mannel, Marcus, et al. "St. John's Wort Extract LI160 for the Treatment of Depression with Atypical Features—A Double-Blind, Randomized, and Placebo-Controlled Trial." *Journal of Psychiatric Research* 44.12 (2010): 760–67.

McFarland, Bentson, et al. "Complementary and Alternative Medicine Use in Canada and the United States." *American Journal of Public Health* 92.10 (2002): 1616–18. Use of CAM in the US and Canada.

Michael, Marilynn. "$34B Spent on Unconventional Care." *National Center for Homeopathy* (31 July 2009). www.homeopathic.org/content/34b-spent-on-unconventional-care. Amount spent on homeopathy.

Milazzo, Stefania, and Edzard Ernst. "Newspaper Coverage of Complementary and Alternative Therapies for Cancer—UK 2002–2004." *Supportive Care in Cancer* 14.9 (2006): 885–89.

Milgrom, Lionel R. "Towards a New Model of the Homeopathic Process

Based on Quantum Field Theory." *Forsch Komplementmed* 13 (2006): 174–83.

Mirowski, Philip, and Robert Van Horn. "The Contract Research Organization and the Commercialization of Scientific Research." *Social Studies of Science* 35 (2005): 530–48.

Moffatt, Barton, and Carl Elliott. "Ghost Marketing: Pharmaceutical Companies and Ghostwritten Journal Articles." *Perspectives in Biology and Medicine* 50 (2007): 18–31.

Moore, Matthew. "Boots Hit by Mass Homeopathy 'Overdose.'" *Telegraph* 19 January 2010.

Moynihan, Ray, Iona Heath, and David Henry. "Selling Sickness: The Pharmaceutical Industry and Disease Mongering." *BMJ* 324 (2002): 886–91.

"OECD Health Care Data." Organisation for Economic Co-operation and Development (June 2005). www.oecd.org/dataoecd/35/13/34966969.pdf.

Papanikolaou, George N., et al. "Reporting of Conflicts of Interest in Guidelines of Preventive and Therapeutic Interventions." *BMC Medical Research Methodology* 1 (2001): 3.

Petsko, Gregory A. "When the Pie Is Too Small." *Genome Biology* 11 (2010): 127. Comments on size of NIH budget.

"Placebo Effect Measured in Spine." CBC News 15 October 2009. www.cbc.ca/news/health/story/2009/10/15/placebo-effect-spinal-cord.html.

Poling, Samantha. "Doctors Warn over Homeopathic 'Vaccines.'" BBC News Scotland 13 September 2010. www.bbc.co.uk/news/uk-scotland-11277990.

Prosser, Helen, Solomon Almond, and Tom Walley. "Influences on GPs' Decisions to Prescribe New Drugs—The Importance of Who Says What." *Family Practice* 20 (2003): 61–68.

"Research Project Success Rates by NIH Institute for 2009." National Institutes of Health (2009). http://report.nih.gov/award/success/Success_ByIC.cfm.

Robbins, Martin. "Homeopathy and Dr. James Le Fanu: If This Is a Witch Hunt, Help Me Find a Torch." *Telegraph* 7 July 2010.

Robert Schad Naturopathic Clinic. The Canadian College of Naturopathic Medicine. www.ccnm.edu/naturopathic_medicine_rsnc.

Ross, Joseph S., et al. "Guest Authorship and Ghostwriting in Publications Related to Rofecoxib: A Case Study of Industry Documents from Rofecoxib Litigation." *Journal of the American Medical Association* 299 (2008): 1800–12.

Schmidt, Katja, and Edzard Ernst. "Assessing Websites on Complementary and Alternative Medicine for Cancer." *Annals of Oncology* 15 (2004): 733–42.

Sinnema, Jodie. "Patients Left Bruised, but Grateful: Say Chiropractic Tool Technique Brings Relief." *Edmonton Journal* 16 August 2010.

Sirois, Fuschia M. "Motivations for Consulting Complementary and Alternative Medicine Practitioners: A Comparison of Consumers from 1997–8 and 2005." *BMC Complementary and Alternative Medicine* 8 (2008): 16.

Sismondo, Sergio. "Ghost Management: How Much of the Medical Literature Is Shaped behind the Scenes by the Pharmaceutical Industry?" *PLoS Medicine* 4 (2007). www.plosmedicine.org/article/info:doi/10.1371/journal.pmed.0040286. Quote from Pfizer sales document regarding the industry purpose of scientific data.

Spurling, Geoffrey K., et al. "Information from Pharmaceutical Companies and the Quality, Quantity, and Cost of Physicians' Prescribing: A Systematic Review." *PLoS Medicine* 7.10 (2010). www.plosmedicine.org/article/info%3Adoi%2F10.1371%2Fjournal.pmed.1000352.

Stelfox, Henry T., et al. "Conflict of Interest in the Debate over Calcium-Channel Antagonists." *New England Journal of Medicine* 2 (1998): 101–6.

Suarez-Almazor, Maria E., et al. "A Randomized Controlled Trial of Acupuncture for Osteoarthritis of the Knee: Effects of Patient-Provider Communication." *Arthritis Care and Research* 62.9 (2010): 1229–36.

Taylor, Paul. "Health Care, under the Influence." *Globe and Mail* 26 April 2008.

Taylor, Rosie, and Jim Giles. "Cash Interests Taint Drug Advice." *Nature* 437 (2005): 1070–71.

Tilburt, Jon C., et al. "Alternative Medicine Research in Clinical Practice: A U.S. National Survey." *Archives of Internal Medicine* 169.7 (2009): 670–77.

Tobin, Martin. "Assessing the Performance of a Medical Journal." *American Journal of Respiratory and Critical Care Medicine* 169 (2004): 1268–72.

Turner, Erick, et al. "Selective Publication of Antidepressant Trials and Its Influence on Apparent Efficacy." *New England Journal of Medicine* 358 (2008): 252–60.

Wager, Elizabeth. "A Concern That Drug Companies Cannot Ignore." *Journal of the Royal Society of Medicine* 98 (2005): 448–50.

Walji, Muhammad, et al. "Efficacy of Quality Criteria to Identify Potentially Harmful Information: A Cross-sectional Survey of Complementary and Alternative Medicine Web Sites." *Journal of Medical Internet Research* 6.2 (2004): e21. Potentially harmful information found on websites.

Wardle, Jon. *Regulation of Complementary Medicines: A Brief Report on the Regulation and Potential Role of Complementary Medicines in Australia*. Brisbane: The Naturopathy Foundation, 2009. CAM use increasing in Australia.

Wazana, Ashley. "Physicians and the Pharmaceutical Industry: Is a Gift Ever Just a Gift?" *Journal of the American Medical Association* 283 (2000): 373–80.

Weeks, Laura, Maria Verhoef, and Catherine Scott. "Presenting the Alternative: Cancer and Complementary and Alternative Medicine in the Canadian Print Media." *Supportive Care in Cancer* 15.8 (2007): 931–38.

Whorton, James C. *Nature Cures: The History of Alternative Medicine in America*. New York: Oxford University Press, 2002. Used for the history of CAM, especially naturopathy.

Wilkinson, Michael H.F. "A Quantum Mechanical Interpretation of Homeopathy." *Annals of Improbable Research* 5.4 (1999): 21–23.

Yaphe, John, et al. "The Association between Funding by Commercial Interests and Study Outcome in Randomized Controlled Drug Trials." *Family Practice* 18 (2001): 565–68.

CHAPTER 5: MAGIC

Angell, Marcia. *The Truth about Drug Companies: How They Deceive Us and What to Do about It.* New York: Random House, 2004.

Bubela, Tania, and Timothy Caulfield. "Role and Reality: Technology Transfer at Canadian Universities." *Trends in Biotechnology* 28 (2010): 447–51. A review of the commercialization trend in Canada.

DeAngelis, Catherine. "Conflict of Interest and the Public Trust." *Journal of the American Medical Association* 284 (2000): 2237–38.

Doucet, Mathieu, and Sergio Sismondo. "Evaluating Solutions to Sponsorship Bias." *Journal of Medical Ethics* 34 (2008): 627–30.

"Evidence-Based Health Care." The Cochrane Collaboration. www.cochrane.org/about-us/evidence-based-health-care.

Henry, David. "Doctors and Drug Companies: Still Cozy after All These Years." *PLoS Medicine* 7.11 (2010): e1000359.

Lemmens, Trudo. "Leopards in the Temple: Restoring Scientific Integrity to the Commercialized Research Scene." *Journal of Law Medicine and Ethics* 32.4 (2004): 641–57. The value of an independent clinical trial research system.

Lewis, Tracy R., Jerome H. Reichman, and Anthony D. So. "The Case for Public Funding and Public Oversight of Clinical Trials." *The Economists' Voice* 4.1 (2007). http://works.bepress.com/tracy_lewis/2.

Pollara Research and Earnscliffe Research and Communications. *Public Opinion Research in Biotechnology Issues—Third Wave.* Ottawa: Earnscliffe Research and Communications, 2000. Contains survey about trust.

Popper, Karl R. *The Logic of Scientific Discovery.* New York: Basic Books, 1959.

Steinbrook, Robert. "Controlling Conflict of Interest—Proposals from the Institute of Medicine." *New England Journal of Medicine* 360 (2009): 2160–63. Review of institute's recommendations for handling conflict of interest in research.

Willett, Walter. "The Mediterranean Diet: Science and Practice." *Public Health Nutrition* 9 (2006): 105–10. References the value of diet and exercise.

Young, Neal S., John P.A. Ioannidis, and Omar Al-Ubaydli. "Why Current Publication Practices May Distort Science." *PLoS Medicine* 5.10 (2008): e201.

ACKNOWLEDGMENTS

This book sprang out of a combined obsession with "facts" (a trait I inherited from my mother), a personal interest in health and fitness, and two decades of involvement with science and health policy. I have been blessed with a job that I enjoy so much that it seems entirely inappropriate to call it a job. And the reason I love my work is that it provides an excuse to interact with interesting people, learn about fascinating new research, and work on challenging issues. This book was an opportunity to do all those things on a singularly important topic: health.

I have so many people to thank for making this book happen. But my first big thanks must go to my good friend and writing mentor Curtis Gillespie. Without his support and nagging nudges, this book would not have happened. His valuable input and editorial advice had an impact on every page of the manuscript. He sets a high standard, both as a writer and as a friend. I am also deeply indebted to my whole research team at the University of Alberta, including Nina Hawkins, Amy Zarzeczny, Nola Ries, Jacob Shelley, Ubaka Ogbogu, Christen Rachul, and Zubin Master. A special thanks to my colleague Robyn Hyde-Lay, who read early drafts and provided a constant stream of useful research material. In addition, all the students I am privileged to work with are a constant source of inspiration, and many—including Lindsey Jo Erhman, who worked on the naturopathic website project, and the Murdoch brothers, CJ and Blake, who did loads of background research—helped with specific projects that were relevant to the book.

Books of this nature require access to and the cooperation of innumerable interviewees. Without exception, all put up with my clumsy questions and enthusiastically guided me through their areas of expertise. It was a true privilege to get to know so many brilliant researchers and health experts. I can't thank them enough for taking the time to help me with this project. You will note that I attempted to interview both well-known senior scholars and new emerging talent (fresh perspectives are key!). But regardless of the stage of their careers, all interviewees were kind, patient, and helpful. This includes the alternative practitioners I visited for the remedies chapter. Though we may not agree about the value of alternative medicine, I found these individuals to be tremendously professional and conscientious. Many of the people I interviewed did not make it into the text of the book, but their insights still played an important role. These individuals include some close friends and, I hope, some new ones, such as Heather Boon (University of Toronto), Trudo Lemmens (University of Toronto), Dana Olstad (University of Alberta), Ross Upshur (University of Toronto), Matthew Nisbet (American University), Wylie Burke (University of Washington), Michael McDonald (UBC), John Spence (University of Alberta), Mildred Cho (Stanford), Scott Roberts (University of Michigan), Daryl Pullman (Memorial University), Neal Cohen (UCSF), Barbara Koenig (Mayo Clinic), Eric Meslin (Indiana University), Amy McGuire (Baylor), Ron Ackerman (Indiana University), George Church (Harvard), Bartha Knoppers (McGill), Stephen Minger (King's College London), and Jeff Woods (a personal trainer). Numerous individuals also provided resources and email assistance, including Jennifer Smith Maguire (University of Leicester) on the nature of the fitness industry; David Colquhoun (University College London) on alternative medicine; Judy Chepeha (University of Alberta) on therapeutic massage and the Graston technique; and Steven N. Blair (University of South Carolina), Bruce Reeder (University of Saskatchewan), and Antero Kesaniemi (University of Oulu) on stretching and flexibility.

I am also indebted to my many wonderful colleagues at the University of Alberta, especially those who suffered through my frequent and annoyingly animated hallway lectures on the latest finding in the field of health science. Many of these individuals, such as Tania Bubela, Ted DeCoste, Wayne Renke, Erin Nelson, and Camilla Knight, sent me useful literature. I have the good fortune to work with countless other amazing colleagues throughout the world. They are too numerous to mention, but know that I value their inspiring collaborations and constant support.

I am thankful to the many funding agencies that have supported my research, including the Canada Research Chair program, the Alberta Heritage Foundation for Medical Research, the Stem Cell Network (National Centres of Excellence), AllerGen (National Centres of Excellence), Genome Alberta, the Canadian Institutes of Health Research, the Social Sciences and Humanities Research Council, the Alberta Policy Coalition for Cancer Prevention, and the Alberta Law Foundation.

My wonderful agents, Anne McDermid and Chris Bucci, have provided unflagging encouragement. Their always practical advice helped me to focus my work. I am also greatly indebted to Diane Turbide at Penguin Canada and Helene Atwan at Beacon Press for taking a chance on this book and on me. And to Jonathan Webb for his exceptional editorial work. Jonathan was firm when he needed to be and always kind, even though I deserved less. I know he had to endure many cringe-worthy typos, unintended logic loops, and mangled phrases.

My entire family—specifically the Caulfield and Otto clans—was an important part of this book. They participated in many of my ridiculous "experiments" and allowed me to record their reactions to various stages of my research. They are a remarkable collection of individuals, good sports all. My brothers Case and Sean provided both specific advice on early drafts and, more importantly, a lifetime of intellectual inspiration.

But my deepest thanks must go to my amazing kids, Adam, Alison, Jane, and Michael, for putting up with their crazy dad, and to my wife, Joanne. Joanne's constant support, insightful (and always correct) advice, and remarkable patience made this book a reality.

INDEX